SpringerBriefs in Health C
and Economics

For further volumes:
http://www.springer.com/series/10293

Alexander Kolker

Healthcare Management Engineering: What Does This Fancy Term Really Mean?

The Use of Operations Management
Methodology for Quantitative
Decision-Making in Healthcare Settings

 Springer

Alexander Kolker
Children's Hospital and Health System
Milwaukee, WI 53201, USA
alexanderkolker@yahoo.com

ISBN 978-1-4614-2067-5   e-ISBN 978-1-4614-2068-2
DOI 10.1007/978-1-4614-2068-2
Springer New York Dordrecht Heidelberg London

Library of Congress Control Number: 2011941803

Printed on acid-free paper

Springer is part of Springer Science+Business Media (www.springer.com)

*You cannot do today's job with yesterday's
methods and be in business tomorrow.*

Anonymous

*Hunch and intuitive impressions are
essential for getting the work started,
but it is only through the quality of the
numbers at the end that the truth can be told.*

S. Glantz, Primer of Biostatistics, 2005.
McGraw-Hill/L. Thomas, Biostatistics in
Medicine. Science, 198: 675, 1977

Preface

This Brief Series book illustrates in depth a concept of healthcare management engineering and its domain for hospital and clinic operations. Predictive and analytic decision-making power of management engineering methodology is systematically compared to traditional management reasoning by applying both side by side to analyze 26 concrete operational management problems adapted from hospital and clinic practice. The problem types include: clinic, bed, and operating rooms capacity; patient flow; staffing and scheduling; resource allocation and optimization; forecasting of patient volumes and seasonal variability; business intelligence and data mining; and game theory application for allocating cost savings between cooperating providers.

Detailed examples of applications are provided for quantitative methods such as discrete event simulation, queuing analytic theory, linear and probabilistic optimization, forecasting of a time series, principal component decomposition of a dataset and cluster analysis, and the Shapley value for fair gain sharing between cooperating participants. A summary of some fundamental management engineering principles is provided.

The goal of the book is to help to bridge the gap in mutual understanding and communication between management engineering professionals and hospital and clinic administrators.

The book is intended primarily for hospital/clinic leadership who are in charge of making managerial decisions. This book can also serve as a compendium of introductory problems/projects for graduate students in Healthcare Management and Administration, as well as for MBA programs with an emphasis in Health care.

What Is This Book About?

Modern medicine has achieved great progress in treating individual patients. This progress is based mainly on life science (molecular genetics, biophysics, and biochemistry) and the development of medical devices, drugs, and imaging technology.

However, according to the highly publicized report "*Building a Better Delivery System: A New Engineering/Healthcare Partnership*" published jointly by the National Academy of Engineering and Institute of Medicine, relatively little material resources and technical talent have been devoted to the proper functioning of the overall healthcare delivery as an integrated system in which access to efficient care is delivered to many thousands of patients in an economically sustainable way (Reid et al. 2005).

A system is generally defined as a set of interconnected elements-subsystems (objects and/or people) that form a complex whole that behaves in ways that these elements acting independently would not. The system boundaries can be defined at different levels (scales). For example, a healthcare system can be defined at the nationwide level; in this case, the main interdependent and connected elements of the system are separate hospitals and large clinics and/or their networks, insurance companies, government bodies, such as the center for Medicaid/Medicare services (CMS), etc.

At a lower level, a system can be defined as a stand-alone hospital; in this case the main interdependent and connected elements of the system are hospital departments, such as emergency, surgical, intensive care, etc.

Management engineering methodology can be applied at all system levels (scales). However, specific method can be different depending on the system scale and complexity. For example, system dynamics that operates mostly with macrolevel patient volumes, large-scale patient categories, and large financial flows and allocations can be an appropriate method to analyzing the nationwide healthcare system and analysis of policy issues. On the other hand, such a powerful methodology as discrete event simulation that operates mostly with individual patients as entities can be more appropriate to analyzing operations at lower scale systems such as a separate hospital. At the same time, the separate hospital, despite a lower scale level, is a complex system in itself, comprised of many interdependent departments and units.

The focus of this book is the application of management engineering principles and methodologies on the scale of a separate stand-alone hospital or a large clinic.

The report (Reid et al. 2005) referenced above provides strong convincing arguments that a real impact on quality, efficiency, and sustainability of the healthcare system can be achieved by the systematic and widespread use of methods and principles of system engineering or healthcare management engineering. Lawrence (2010) strongly supports this view, "The opportunities for improvement are substantial; the techniques of industrial engineering can drive large improvements in efficiency and quality and safety when applied within the right context...," and further, "...As we improve performance, innovations will continue around the sick care system, ... in turn enabling the tools of industrial design to have their greatest impact."

Thus, the scope of healthcare management engineering field can broadly be defined as a systematic way of developing managerial decisions for efficient allocating of material, human, and financial resources needed for delivery of high-quality care using the various mathematical and computer simulation methods. (The term "management engineering" is sometimes substituted by the terms "operations research," "system engineering," "industrial engineering," "operations management", or "management science." All these terms have a similar meaning).

Management engineering methodology has become indispensable in addressing *pressing* hospital issues, such as:

- Capacity: How many beds are required for a department or unit? How many procedure rooms, operating rooms or pieces of equipment are needed for different services?
- Staffing: How many nurses, physicians, and other providers are needed for a particular shift in a unit (department) in order to best achieve operational and service performance objectives?
- Scheduling: What are the optimized staff schedules that help not only in delivering a safe and efficient care for patients, but also take into account staff preferences and convenience?
- Patient flow: What patient wait time is acceptable (if any at all) at the service stations in order to achieve the system throughput goals?
- Resource allocation: Is it more efficient to use specialized resources or pooled (interchangeable) resources (operating/procedure rooms, beds, equipment, and staff)? Does it make economic sense to keep some patient service lines (or drop them at all)?
- Forecasting: How to forecast the future patient volumes (demand) or transaction volumes for short- and long-term budget planning and other planning purposes?
- Optimized geographic location of facilities and facilities layout.
- Design of the facility optimized workflow.
- Defining and measuring staff productivity.
- Optimizing a supply chain and inventory management.
- Extracting useful information from raw data sets for marketing and budget planning using Business Intelligence and Data Mining.

This list can easily be extended to include any other area of operational management that requires quantitative analysis to justify decision-making.

The ultimate goal and the holy grail of management engineering methodology is to provide an aid and guidance to efficiently managing hospital operations, i.e., reducing the costs of using resources for delivery of care while keeping high safety and outcome standards for patients.

It can be said that the entire hospital is a patient; management engineering methodology is a medical field with different specialties developed to address different conditions and operational problems; and the management engineer serves as a doctor who diagnoses the operational disease and develops a treatment plan for an ailing hospital and its operations.

No concept or methodology can truly be convincing without multiple concrete and practically relevant examples of its application. Kopach-Konrad et al. (2007) state, "…we believe that widespread success will only come when a critical mass of health care organizations recognize its [healthcare engineering] value through concrete examples. Only then will these organizations promote changes needed for its adoption." The author of this book shares this belief.

In this book traditional managerial decision-making and management engineering methodology are applied side by side to analyze 26 concrete operational management problems adapted from a hospital and clinic practice. The focus is on explaining why management engineering results are often different from the typical, traditional "common sense" management approach. The problems included in this book are somewhat simplified and adapted to focus on the fundamental principles of quantitative decision-making. However, even simplified, most of the problems are not trivial.

It is not possible to illustrate all of the above-mentioned management engineering applications in one Brief Series book. This book covers detailed examples of five types of problems.

The first type is one of the most practically important and widespread problems of dynamic supply and demand balance. This includes 11 capacity and patient flow problems and 5 staffing and scheduling problems. It is widely acknowledged that the most powerful and versatile methodology for analyzing these kinds of problems is discrete event simulation (DES).

At the same time, queuing analytic theory (QAT) is often recommended as a means of analyzing hospital capacity, patient flow, and staffing issues (Litvak 2007; McManus et al. 2004; Haraden et al. 2003). However, such a recommendation underestimates some serious practical limitations of QA theory for hospital applications (D'Alesandro 2008) but overestimates difficulties of using DES. These two methodologies (QAT and DES) are applied to the same problems (along with a traditional managerial approach) to demonstrate side by side the pros and cons of each.

Discrete event simulation software package used to develop example models in this book is ProcessModel 5.3.0 (Process Model, Inc., http://www.processmodel. com). However practically any other high-level commercially available DES package can also be used, such as ProModel, Arena, Simul8, AnyLogic, Simio, FlexSim and many others. All provide a user-friendly graphical interface that makes the efforts of building realistic simulation models no more demanding than the efforts to make simplifications, adjustments, and calibrations to develop rather complex but limited analytic queuing models.

Swain (2007), Abu-Taieh and El Sheikh (2007), Hlupic (2000), and Nikoukaran (1999) provided a review and a comparative study of dozens of commercially available discrete event simulation packages.

The second type of problem includes five problems for the linear and probabilistic resource optimization and allocation. This section includes linear optimization of patient service volumes for different service lines, optimal staffing for 24/7 three-shift operations, physician resident scheduling to meet Institute of Medicine (IOM) new restricted resident work hours for day and night shifts, optimized specimen mass screening testing aimed at reducing the overall number of tests per specimen, and the projection of the expected number of patients discharged from the Emergency Department given the time that a patient has already stayed in ED (the use of the concept of the conditional probability of discharge).

The third type of problem includes two problems for forecasting of a time series using past data points. It is argued that the past data points used for forecasting the future data points should be strongly correlated to each other. It is illustrated that the strongly correlated past data points can be identified from the autocorrelation function of the time series. It is further illustrated that a powerful forecasting procedure for the time series can be a recursive technique. Its application is demonstrated using, as examples, annual patient volume forecasting, as well as forecasting of the seasonal variation of the hemoglobin A1C level.

The fourth type of problem includes an application of business intelligence (BI) and data mining (DM) based on advanced multivariate data analysis. Two examples are presented: principal component decomposition of the large dataset of original observational data to identify contributing independent variables (BI), and cluster analysis (DM). Both principal component decomposition and cluster analysis are techniques for data volume reduction.

Principal component decomposition is used to reduce the number of variables (the number of columns) in observational datasets. Because of inevitable intercorrelation of some variables in large observational datasets these variables carry little or no independent information. These variables are redundant. Therefore identifying contributing independent variables (factors) using regression analysis with dozens of the original variables (including redundant) usually fails. Principal component decomposition technique allows one to identify those redundant variables and retain only a few mutually uncorrelated principal variables (components). This makes possible to identify a few independent contributing factors using regression analysis with only a rather small number of totally uncorrelated principal components that retain practically all original information. This technique is one of the most powerful for identifying only a few significant independent contributing variables (factors).

Cluster analysis is used to reduce the number of observations of the variables (the number of rows of this data set). Cluster analysis is widely used to identify the groups of objects (clusters) in such a way that the objects within one cluster have much higher similarity compared to objects within another cluster; thus, the total number of observational values of the variable is replaced by a few clusters of observational values. Thus, if there is some known information about an object, one can take full advantage of this information to understand the other objects in the cluster to which this object belongs.

The fifth type of problem is an application of mathematical game theory for allocating cost savings (gains) between cooperating participants. The problem of cost allocation between parties arises in accounting of practically every organization. As an alternative to traditional accounting allocation basis, there is a growing interest in cost allocation principles based on logic of game theory. The most widely used method of joint-cost (savings) allocation is the Shapley value. The Shapley value method provides a fair cost allocation in which each participant estimates the net benefits (net values) expected from cooperation with other groups of participants (coalitions).

In conclusion, some fundamental management engineering principles for efficient managerial decision-making in healthcare settings are summarized. Other quantitative methods that can be used for various management engineering applications are summarized in the appendix.

It is illustrated how truly efficient managerial decisions can be developed and why traditional management approaches usually result in unsatisfactory and short-lived outcomes.

Who Is This Book For?

The already referenced report published by National Academy of Engineering and Institute of Medicine (Reid et al. 2005) states in an unusually blunt way, "In fact, relatively few health care professionals or administrators are equipped to think analytically about health care delivery as a system or to appreciate the relevance of engineering tools. Even fewer are equipped to work with engineers to apply these tools."

Thus, it is often difficult for many administrators to appreciate the role of management engineering methodology in the healthcare delivery process analysis. On the other hand, engineering professionals do not always have enough knowledge of healthcare delivery processes or the role of physicians in making not only clinical but also management decisions. Health care has a culture of rigid division of labor. This functional division does not effectively support the methodology that crosses the functional areas, especially if it assumes significant change in traditional relationships.

As Butler (1995) states, "...it is imperative for administrators to familiarize themselves with the array of quantitative decision techniques provided by management science/operations research (MS/OR)." Vissers (1998) argues, "Modeling-based health care management ought to become just as popular as evidence-based medicine. Making managerial decisions based on evidence by modeling efforts is certainly a step forward." Carter (2002) in the article with the revealing title 'Diagnosis: Mismanagement of Resources' summarizes, "...Ailing health care system desperately needs a dose of operations research." Fabri (2008) supports this assessment saying, "...fixing healthcare will require individuals who are 'bilingual' in healthcare and in systems engineering principles." Kopach-Konrad et al. (2007) also support this view, "...medical professionals and managers need to understand and appreciate the power that systems engineering concepts and tools can bring to re-designing and improving health care environments and practices."

Berwick (2011), the currently appointed Administrator of the Centers for Medicare and Medicaid Services (CMS) and the former President of the Institute for Healthcare Improvement (IHI), writes in the same line in the recently published proceedings of the workshop "Engineering a Learning Healthcare System: A Look at the Future" (IOM 2011), "Healthcare leaders tend not to be aware of the engineering disciplines or to be suspicious of their applicability... Bridge building here will be expensive, and it will take time, but it will pay off."

Valdez and Brennan (2009) presented a background report that provides a critical summary of 13 other reports sponsored by the various national bodies (National Academy of Science, National Academy of Engineering, Institute of Medicine, and National Science Foundation). This report explores the issues at the intersection between system engineering and health care. One of the key themes is the opportunities for cross-education and collaboration between healthcare and engineering professionals.

Compton and Reid (2008) argue in the same line, "...most health care professionals do not even know what questions to ask system engineers nor what to do with the answers, and vice versa… Few system engineers understand the constraints under which health care providers operate. In short, these two groups of professionals often talk to each other but seldom understand each other."

Story (2009) states in his strongly articulated article, "Local hospital leadership must play a role, either becoming educated and passionate about systems [management] engineering or stepping aside to allow progress. …Leaders who remain unfamiliar with the concepts of system engineering often slow or even prevent change through a lack of passion, education and vision." While this view might look extreme, it nonetheless reflects the depth of frustration and one of the root causes of the problem with practical implementation of management engineering methodology.

At the same time, for the last few years, some positive signs have been observed as management engineering slowly makes its way into hospital settings. Story (2009) himself notices, "…while we still have much work to do, especially at the hospital level, some change addressing the recommendations for the dissemination of systems engineering and systems thinking is already under way." A similar conclusion on advancing the role of management engineering in healthcare settings is made by Buttell Crane (2007). It is also an encouraging sign that the above-referenced Kopach-Konrad et al. (2007) article appeared not in an engineering or operation research journal or on specialized website, but it was published in such an authoritative medical journal as the Journal of General Internal Medicine.

Another encouraging sign is the already referenced proceedings of the workshop (IOM 2011). This workshop drew together participants from healthcare and engineering disciplines to identify challenges in health care that might benefit from a system engineering perspective.

To address the challenge of transforming the system of care delivery in practice, some leading healthcare organizations have adopted this area as a strategic priority. For example, the Mayo Clinic, one of the largest integrated medical centers in the USA, has defined the Science of Healthcare Delivery as one of its four strategic directions. The others are Quality, Individualized Medicine, and Integration (Fowler et al. 2011). The Mayo Clinic has also created the Center for the Science of Healthcare Delivery, a new initiative that will focus on creating improved approaches to how health care is delivered (Mayo Clinic 2011).

Although all of the above references on the role and importance of management (system) engineering for health care are insightful and to the point, most of them do not include detailed practical examples of application of management engineering approach. This makes them somewhat declarative.

On the other hand, for the last 40 years or so, hundreds of journal articles and conference presentations have demonstrated the power and benefits of using management engineering in healthcare settings. However, most are written by university researchers and/or engineering professionals, and are rather complex as a guide for typical hospital administrators who are in charge of making managerial decisions.

Thus, a wide gap exists between the publications that urge the use of management engineering in healthcare settings but provide few or no practical examples, and the publications with examples that are too specialized and complex for digesting by a typical hospital administrator. This gap is probably one of the reasons why too many administrators still have a vague idea of the practical value of healthcare management engineering methodology. Many of them simply do not see "what's in it for me."

The main objective of this book is to contribute to filling this gap. Thus, this book is intended primarily for hospital/clinic leaders who have the power to making managerial decisions (department/clinic directors, Vice-Presidents, Chief Operations Officers, CEOs, board members, among others). At the same time, this book can serve as a concise text and compendium of problems/projects for graduate students in Healthcare Management and Administration, as well as for MBA programs with an emphasis in Health care.

This book does not provide all technical details needed for developing simulation models, making a forecast or performing a complex statistical analysis of the database (it is a separate area of professional expertise and the subject of other books). Rather, this book seeks to help hospital leadership to understand why traditional management approaches are often not accurate, short-lived, or unsustainable; which quantitative technique is more appropriate for addressing a particular managerial problem; what can be expected from a particular technique and what are its strength and limitations. For example, is queuing analytic theory (QAT) or discrete event simulation (DES) appropriate methodology for addressing a particular problem? What are the caveats of using QAT and DES and what type and amount of data is needed for each? What type of problem is linear optimization (LO) or game theory for? What are the caveats in interpretation of LO or cluster analysis results? What technique is the most appropriate for making a particular forecast type and why? What is the best approach to the fair cost (savings) allocation? And so on…

In other words, this book seeks to demonstrate to orchestra conductors a role and sound of particular musical instruments rather than to provide techniques for professionally playing the particular instrument (mastering such techniques would require special training similar to physician resident training).

The author sincerely hopes that the multiple examples included in this book would help to bridge the gap in mutual understanding and communication between management engineering professionals and hospital administrators.

Milwaukee, WI, USA   Alexander Kolker

Acknowledgments

The author would like to acknowledge the support and encouragement of Neil Levine, Editor, Operations Research & Management Science, and Matthew Amboy, Assistant Editor, Business & Economics: Operation Research & Management Science, Springer Science & Business Media. Their help and patience have been indispensable.

No book can be written without the support, encouragement, and patience of loved ones. The author would like to express his indebtedness to his wife, Rosa, and to his daughter, Julia Kolker, for their moral support and understanding during the long evenings and many weekends spent working on this book.

Alexander Kolker

Contents

Chapter 1
Traditional Management and Management Engineering

Abstract Definitions and comparison of traditional management and management engineering approaches are provided. Factors that contribute to the difference in these approaches are discussed.

Keywords Traditional management • Management decision • Variability • Uncertainty • Interdependence • Scaling effect

There are many possible definitions of management. For the purpose of this book, we define management as controlling and leveraging available resources (material, financial, and human) aimed at achieving system performance objectives.

Traditional healthcare management is based on past experience, feelings, intuition, educated guesses, simple linear projections, and calculations based on the average values of input variables.

In contrast, management engineering is the discipline of building mathematical models of real systems and analysis thereof for the purpose of developing justified managerial decisions. Management decisions for leveraging resources that best meet system performance objectives are based on comparative outcomes of validated mathematical models.

Although no formal definition can capture all aspects of the concept, it follows that management engineering typically includes the following elements (steps): (1) the goal that is clearly stated and measurable, (2) identification of available resources that can be leveraged (allocated) in different ways, and (3) mathematical models (analytic or numeric computer algorithms) to quantitatively test outcomes (scenarios) for different ways of using resources, and consequences (especially unintended consequences) of the different use of resources before finalizing the decisions.

As Joustra et al. (2011) note, "Currently, management lacks the proper decision support for determining the consequences of their decisions and therefore for making good choices." The underlying foundation of the management engineering approach to quantitative managerial decision-making is that analysis of a valid mathematical model forms a valid basis for truly justified managerial decisions.

Decisions based on management engineering methodology are often different compared to traditional management decisions. Sometimes, they even look counter-intuitive. There are several factors that contribute to this difference.

First, most managerial decisions in healthcare settings are being made in highly variable and random environments. It is a general human tendency to avoid the complications of incorporating uncertainty and randomness into decision-making by ignoring it or turning it into certainty. For example, the average procedure time or the average patient length of stay or the average number of patients are typically treated as if they are fixed values, ignoring the effect of variability around these averages. This practice usually results in seriously inaccurate conclusions made by traditional management decision-making.

Another factor is that healthcare systems usually contain internal hidden inter-connections and interdependencies of units, departments, physicians, nursing and other staffing, regulators, and so on. These multiple interconnections make health-care systems truly complex. Traditional management lacks a means of capturing such interconnections and predicting their effect on the response of one unit to the change in other units. After all, a hospital or a large clinic looks more like, say, a chaotic, busy airport than an automated manufacturing assembly line. This is a root cause of the frequently observed unintended and undesired consequences of mana-gerial decisions that look reasonable on the surface.

One more factor that contributes to the difference between traditional manage-ment and management engineering decision-making is a nonlinear scaling effect (size effect) of most healthcare systems. Larger systems can function at a much higher utilization level and lower patient waiting time than smaller systems even if the patient volume relative to their size is the same (Green 2006, Kolker 2011). Such nonlinear relationships are not easy to incorporate into traditional decision-making.

Only analytic models (when applicable) or computer simulation models offer a means of capturing all these factors into the efficient managerial decision-making.

Chapter 2
Dynamic Supply and Demand Balance Problems

Abstract Comparative analysis of traditional approach, queuing analytic theory formulas (QAT), and discrete event simulation (DES) is provided. Eleven capacity and patient flow problems and five staffing and scheduling problems are analyzed in details using side-by-side a traditional approach, QAT and DES. Serious limitations of QAT compared to DES are demonstrated.

Multiple examples of inaccurate decisions based on input of the average data are illustrated (the flaw of averages).

Keywords Discrete event simulation • Queuing analytic theory • Flaw of averages • Patient flow • Interdependency • Staffing • Variability

2.1 Discrete Event Simulation Methodology: What Is a Discrete Event Simulation Model and How Does a Simple Model Work?

A discrete event simulation (DES) model of a system/process is a computer model that mimics the dynamic behavior of the system/process as it evolves with time in order to visualize and quantitatively analyze its performance. The validated and verified model is then used to study the behavior of the original system/process and its response to input variables in order to identify the ways for its improvement (scenarios) based on some improvement criteria. This strategy is significantly different from the hypothesis-based testing widely used in medical research (Kopach-Konrad et al. 2007).

DES models track entities moving through the system at distinct points of time (events). The detailed track is recorded for all processing times and waiting times. Then the system's statistics for entities and activities are gathered.

To illustrate how a DES model works step by step, let us consider a very simple system that consists of a single patient arrival line and a single server. Suppose that patient interarrival time is random and uniformly (equally likely) distributed between

A. Kolker, *Healthcare Management Engineering: What Does This Fancy Term Really Mean?*, SpringerBriefs in Health Care Management and Economics, DOI 10.1007/978-1-4614-2068-2_2, © Alexander Kolker 2012

1 and 3 min. Service time is also random and distributed exponentially with the average 2.5 min (Of course, any statistical distributions or nonrandom patterns can be used instead). A few random numbers sampled from these two distributions are, for example:

Interarrival time (min)	Service time (min)
2.6	1.4
2.2	8.8
1.4	9.1
2.4	1.8

Let us start our example simulation at time zero, $t=0$, with no patients in the system. We will be tracking any change or event that happened in the system.

A summary of what is happening in the system looks like this:

Event #	Time, min	Event that happened in the system
1	2.6	First customer arrives. Service starts that should end at time $=4$
2	4	Service ends. Server waits for patient
3	4.8	Second patient arrives. Service starts that should end at time $= 13.6$. Server idles 0.8 min
4	6.2	Third patient arrives. Joins the queue waiting for service
5	8.6	Fourth patient arrives. Joins the queue waiting for service
6	13.6	Second patient (from event 3) service ends. Third patient at the head of the queue (first in, first out) starts service that should end at time 22.7
7	22.7	Patient #4 starts service...and so on

These simple but tedious logical and numerical event-tracking operations (algorithm) are suitable, of course, only for a computer. However, they illustrate the basic principles of a typical DES model, in which discrete events (changes) in the system are tracked when they occur over the time. In this particular example, we were tracking events at discrete points in time $t=2.6$, 4.0, 4.8, 6.2, 8.6, 13.6, and 22.7 min.

Once the simulation is completed for any length of time, another set of random numbers from the same distributions is generated, and the procedure (called replication) is repeated. Usually multiple replications are needed to properly capture the system's variability. In the end, the system's output statistics is calculated, e.g., the average patient and server waiting time, its standard deviation, the average number of patients in the queue, the confidence intervals, and so on.

In this example, only two patients out of four waited in the queue. Patient 3 waited $13.6-6.2=7.4$ min and patient 4 waited $22.7-8.6=14.1$ min, so the simple average waiting time for all four patients is $(0+0+7.4+14.1)/4=5.4$ min. Notice, however, that the first two patients did not wait at all while patient 4 waited 2.6 times longer than the average. This illustrates that the simple average could be rather misleading as a performance metric for highly variable processes without some additional information about the spread of data around the average.

Similarly, the simple arithmetic average of the number of waiting patients (average queue length) is 0.5. However, a more informative metric of the queue length is the time-weighted average that takes into account the length of time each patient was in the queue. In this case, it is $(1 \times 7.4 + 1 \times 14.1)/22.7 = 0.95$. Usually the time-weighted average is a better system performance metric than the simple average.

DES models are capable of tracking hundreds of individual entities arriving randomly or in a complex pattern, each with its own unique attributes, enabling one to simulate the most complex systems with interacting events and component interdependencies.

Typical DES applications include: staff and production scheduling, capacity planning, cycle time and cost reduction, throughput capability, resources and activities utilization, bottleneck finding, and analysis. DES is the most effective tool to perform quantitative "what-if" analysis and plays different scenarios of the process behavior as its parameters change with time. This simulation capability allows one to make experiments on the computer, and to test different options before going to the hospital floor for actual implementation.

The basic elements (building blocks) of a simulation model are:

- Flow chart of the process, i.e., a diagram that depicts logical flow of a process from its inception to its completion
- Entities, i.e., items to be processed, e.g., patients, documents, customers, etc.
- Activities, i.e., tasks performed on entities, e.g., medical procedures, exams, document approval, customer check-in, etc.
- Resources, i.e., agents used to perform activities and move entities, e.g., service personnel, equipment, nurses, and physicians
- Entity routings that define directions and logical conditions flow for entities

Typical information usually required to populate the model includes:

- Quantity of entities and their arrival time, e.g., periodic, random, scheduled, daily pattern, etc. There is no restriction on the arrival distribution type
- The time that the entities spend in the activities, i.e., service time. This is usually not a fixed time but a statistical distribution. There is no restriction on the distribution type
- Capacity of each activity, i.e., the maximum number of entities that can be processed concurrently in the activity
- The maximum size of input and output queues for the activities
- Resource assignments: their quantity and scheduled shifts

2.2 Queuing Analytic Theory: Its Use and Limitations

The term "queuing theory" is usually used to define a set of analytic techniques in the form of closed mathematical formulas to describe properties of the processes with a random demand and supply (waiting lines or queues). Queuing formulas are usually applied to a number of predetermined simplified models of the real processes for which analytic formulas can be developed.

Weber (2006) writes that "…There are probably 40 (queuing) models based on different queue management goals and service conditions…" and that it is easy "… to apply the wrong model" if one does not have a strong background in operations research.

Development of tractable analytic formulas is possible only if a flow of events in the system is a steady-state Poisson process. On definition, this is an ordinary stochastic process of independent events with the constant parameter equal to the average arrival rate of the corresponding flow. Time intervals between events in a Poisson flow are always exponentially distributed with the average interarrival time that is the inverse Poisson arrival rate. Service time is assumed to follow an exponential distribution or, sometimes, uniform or Erlang distribution. Thus, processes with a Poisson arrival of events and exponential service time are Markov stochastic processes with discrete states and continuous time.

Most widely used queuing models for which relatively simple closed analytical formulas have been developed are specified as *M/M/s* type (Hall 1990; Lawrence and Pasternak 2002). (*M* stands for Markov since Poisson process is a particular case of a stochastic process with no "after-effect" or no memory, known as continuous time Markov process). These models assume an unlimited queue size that is served by *s* providers.

Typically *M/M/s* queuing models allow calculating the following steady-state characteristics:

– Probability that there are zero customers in the system
– Probability that there are *K* customers in the system
– The average number of customers waiting in the queue
– The average time the customers wait in the queue
– The average total time the customer spends in the system ("cycle time")
– Utilization rate of servers, i.e., percentage of time the server is busy

As more complexity is added in the system, the analytic formulas become less and less tractable. Analytic formulas are available that include, for example, limited queue size, customers leaving the system after waiting a specified amount of time, multiple queues with different average service time and different providers' types, different service priorities, etc. However, the use of these cumbersome formulas even built in Excel spreadsheets functions (Ingolfsson and Gallop 2003) or tables (Seelen et al. 1985) is rather limited because they cannot capture complexity of most healthcare systems of practical interest.

Assumptions that allow deriving most queuing formulas are not always valid for many healthcare processes. For example, several patients sometimes arrive in emergency department (ED) at the same time (several people injured in the same auto accident), and/or the probability of new patient arrivals could depend on the previous arrivals when ED is close to its capacity, or the average arrival rate varies during a day, etc. These possibilities alone make the arrival process a nonordinary, nonstationary with after-effect, i.e., a non-Poisson process for which queuing formulas are not valid. Therefore, it is important to properly apply statistical goodness-of-fit tests to verify that the null-hypothesis that actual arrival data follow a Poisson distribution cannot be rejected at some level of significance.

An example of a conclusion from the goodness-of-fit statistical test that is not convincing enough can be found, for instance, in Harrison et al. (2005). The authors tried to justify the use of a Poisson process by using a chi-square goodness-of-fit test. The authors obtained the test p-values in the range from 0.136 to 0.802 for different days of the week. Because p-values were greater than 0.05 level of significance, they failed to reject the null-hypothesis of Poisson distribution (accepted the null-hypothesis).

On the other hand, the fundamental property of a Poisson distribution is that its mean value is equal to its variance (squared standard deviation). However, the authors' own data indicated that the mean value was not even close to the variance for at least 4 days of the week. Thus, the use of a Poisson distribution was not actually convincingly justified for patient arrivals. Apparently, chi-square test p-values were not large enough to accept the null-hypothesis with high-enough confidence (alternatively, the power of the statistical test was likely too low).

Despite its rather limited applicability to many actual patient arrival patterns, a Poisson process is widely used in operation research as a standard theoretical assumption because of its mathematical convenience (Gallivan et al. 2002; Green 2006; McManus et al. 2003).

Some authors are trying to make queuing formulas applicable to real processes by fitting and calibration. For example, in order to use queuing formulas for a rather complex ED system, Mayhew and Smith (2008) made a significant process simplification by presenting the workflow as a series of stages. The stages could include initial triage, diagnostic tests, treatment, and discharge. Some patients experienced only one stage while others more than one. However, the authors acknowledge "… what constitutes a 'stage' is not always clear and can vary…and where one begins and ends may be blurred." The authors assumed a Poisson arrival and exponential service time but then used actual distribution service time for "calibration" purposes. Moreover, they observed that exponential service time for the various stages "…could not be adequately represented by the assumption that the service time distribution parameter was the same for each stage." In the end, all the required calibrations, adjustments, fitting to the actual data made the model to lose its main advantage as a queuing model: its analytical simplicity and transparency. On the other hand, all queuing formulas assumptions and approximations still remained.

Thus, many complex healthcare systems with interactions and interdependencies of the subsystems cannot be effectively analyzed using analytically derived closed formulas.

Moreover, queuing formulas cannot be directly applied if the arrival flow contains a nonrandom component, such as scheduled arrivals (see Sects. 2.3.6–2.3.10). Therefore, in order to use analytic queuing formulas, the nonrandom arrival component should be first eliminated, leaving only random arrival flow for which QA formulas could be used (Litvak 2007).

Green (2004) applied *M/M/s* model to predict delays in the cardiac and thoracic surgery unit with mostly elective scheduled surgical patients assuming a Poisson pattern of their arrivals. The author acknowledged that this assumption could result in an overestimate of delays. In order to justify the use of *M/M/s* model, the author argued that some "…other factors are likely to more than compensate for this." However, it was not clear how much those factors could compensate the overestimated delays.

Still, despite their limitations, queuing analytic theory (QAT) models can be used for application to simply structured steady-state processes if a Poisson arrival and exponential service time assumptions are accurate enough or, at least, a coefficient of variation (the ratio of standard deviation to the mean) is close to 1.

A number of specific examples that compare the use of simple DES and QAT models and illustrate QAT limitations are presented in the next sections.

2.3   Capacity Problems

2.3.1   Outpatient Clinic: Centralized or Separate Locations?

A busy community clinic provides flu shots during a flu season peak on a walk-in basis (no appointment necessary). The clinic is staffed with four nurses. An average patient arrival rate is about 28 patients per h. Giving a shot (including the time for necessary paperwork) takes on average about 8 min.

During a typical clinic operation day, it was observed quite a long waiting line and some patients complained about long waiting time. The clinic operational performance should be improved.

2.3.1.1   Traditional Management Approach

In order to address the issue of long patient waiting time, a brainstorming session was organized. It was decided that for patient convenience and in order to reduce waiting time, the centrally located clinic should be separated to two different more convenient locations, each staffed with two nurses, so the total clinic staffing remains the same. It was assumed that patients will chose a more conveniently located clinic, such that the total patient arrival rate will split about equally between two clinics.

2.3.1.2   Queuing Analytic Theory Application

Instead of brainstorming, the manager decided to apply operations management methodology to evaluate quantitatively an effect of clinic separation. It was assumed that analytic queuing formulas would be applicable in this case.

For the current centralized operation mode, the following *M/M/s* analytical model with the unlimited queue size can be used:

Random patient arrival was assumed to be a Poisson process with the total average arrival rate $\lambda = 28$ patients per h, average flu shot time $\tau = 8$ min, and the number of nurses (servers) $N = 4$.

The final (steady-state) probability that there are no patients in the system, p_0, is calculated using the formula (Hall 1990; Green 2006):

$$p_0 = \left[\sum_{n=0}^{N-1} \frac{a^n}{n!} + \frac{a^N}{N!(1-\rho)} \right]^{-1}, \qquad (2.1)$$

where

$$a = \lambda \times \tau, \quad \rho = a / N.$$

The average number of patients in the queue, L_q, is

$$L_q = \frac{a^{N+1} \times p_0}{[N \times N! \times (1-\rho)^2]} \qquad (2.2)$$

and the average time in the queue, t, is (the Little's formula)

$$t = \frac{L_q}{\lambda}. \qquad (2.3)$$

Substituting in the formulas (2.1)–(2.3) $N=4$, $\lambda=28$ patients per h, and $\tau=8$ min $= 0.1333$ h, we get the average number of patients in the queue $L_q = 11.9$ patients, and the average waiting time about 25.5 min. The clinic's average utilization is 93%.

In practice, an excel spreadsheet is used to perform calculations using queuing formulas.

The new proposed system of two separately located clinics that was supposed to be more convenient for patients and perform better consists of two separate systems with $N=2$. Arrival rate for each separate clinic was going to be $\lambda=28/2=14$ patients per h.

Using formulas (2.1)–(2.3), we get the average number of patients in each clinic queue $L_q = 12.6$. The average waiting time in the queue will be about 54 min!

Thus, in the proposed "improved" clinics the number of patients in each clinic queue will remain about the same while the average waiting time will be about twice of that of the original operations.

It should be concluded that the proposed improvement change that might look reasonable on the surface does not stand the scrutiny of the simple quantitative analysis.

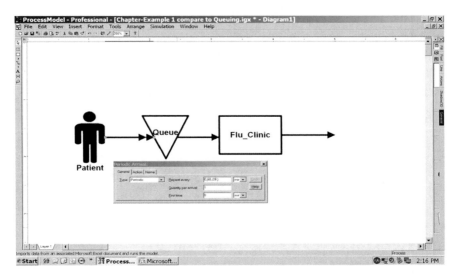

Fig. 2.1 Layout of the simulation model of flu clinic. Information on the panel indicates patient arrival type (periodic) that repeats on average every E(2.143) min

Although the ratio of the number of patients per server remains the same in both cases (28:4 = 7 vs. 14:2 = 7), separating one random patient flow on two equally divided separate random flows with the proportionally divided number of servers does not result in efficiency improvement; it makes things worse. This is because separate servers cannot help each other if some of them become overworked for sometime due to a surge in patient arrival because of random variability of the patient flow.

The use of DES model to analyze the same process without resorting to analytical formulas is given in the next section.

2.3.1.3 Discrete Event Simulation Application

Let us consider the same flu clinic that was analyzed above using QA. DES model layout is presented in Fig. 2.1. It simply depicts the arrived patient flow connected to queue, then coming to the flu clinic (box called Flu_Clinic), and then exit the system. These basic model elements are simply dragged down from the pallet and then connected to each other.

Next step is to fill in the process information: patients arrive periodically, one patient at a random exponentially distributed time interval with the average interarrival time 60 min/28 = 2.143 min, E(2.143), as indicated on the data arrival panel in Fig. 2.1. This corresponds to Poisson arrival rate of 28 patients per h (E stands for exponential distribution). In the Flu_Clinic data panel, the capacity input was 4 (four patients served concurrently by four nurses), and the service time was exponentially random with the average value 8 min, E(8). This completes the model setup.

The model was run 300 replications in order to capture the variability of patient arrivals and service time. The following steady-state (plateau) results are obtained (after the warm-up period of 1,300 h identified by preliminary numeric experimentation with the model).

The average number of patients in the queue is 11.8, and the average waiting time is 25.3 min. Thus, we received practically the same results with the simple DES model as with QA model using no analytic formulas at all.

Similarly, for the separately located clinic with the arrival rate 14 patients per h and two nurses (capacity), the same DES model results in the average steady-state 12.5 patients in the queue and the average waiting time 53.6 min, i.e., again practically the same results as the given above QA formulas.

Thus, both QA and DES models illustrate a fundamental management engineering principle: combined resources with random patient flow and with unlimited queue size and no leaving patients are more efficient than separate resources with the same total work load (see also Sect. 2.3.7). If specialized (dedicated) resources are needed due to patient privacy, infection control, nonmovable equipment or other special factors, then some additional capacity should be planned and budgeted to cover the loss of resources' efficiency. Specialized resources (staff, operating or procedure rooms, beds, etc.) typically cost more than mutually interchangeable (pooled) or shared resources. This principle was also illustrated using cooperative game theory approach in Sect. 6.1.

2.3.2   Outpatient Clinic: Nonsteady-State Operations

The manager of the same clinic decided to verify that the clinic would operate smoothly enough with a new team of less experienced nurses (servers) who would work only a little slower than the previous ones. The average time to give a shot will be about 8.6 min (instead of average 8 min for more experienced staff, as in the previous section).

2.3.2.1   Traditional Management Approach

The manager reasoned that because the difference in time for giving a shot is rather small (only about 0.6 min on average), it would not practically affect the clinic operation: the number of patients in the queue and their waiting time on the typical working day would be practically the same or, at worst, only a little longer than in the previous case.

2.3.2.2   Queuing Analytic Theory Application

In order to verify his/her gut feeling of little difference in clinic performance, the manager plugged the average service time 8.6 min and the arrival rate 28 patients

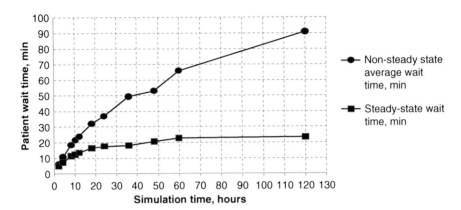

Fig. 2.2 DES simulated average patient wait time. *Upper curve* is nonsteady-state operation (average patient service time is 8.6 min). *Bottom curve* is steady-state operation (average patient service time is 8.0 min)

per h in the *M/M/s* queuing calculator. However, the calculator returned no number at all, which means that no solution could be calculated. Why is that?

An examination of (2.1) shows that if ρ becomes greater than 1, the last term in the sum becomes negative and the calculated p_0 also becomes negative, which does not make sense (If ρ is equal to 1, the term becomes uncertain and the calculations cannot be carried out at all). The average service time in this example is only slightly higher than it was in the previous case. However, this small difference made parameter ρ greater than 1 ($\rho = 1.0033 > 1$). This explains why the calculations cannot be done using this value.

Most queuing analytic formulas are applicable only for steady-state processes, i.e., for the established processes whose characteristics do not change with time. The steady-state condition is possible only if $\rho < 1$; otherwise the queue grows indefinitely. (Queuing nonsteady-state processes are typically described by differential equations, and require rather complex numerical procedures to get solutions.)

2.3.2.3   Discrete Event Simulation Application

In contrast, DES methodology easily handles a nonsteady-state situation and explicitly demonstrates the growth of the queue. Using the same DES model described in the previous section, we simply plug the average service time 8.6 min in the Flu_ Clinic data panel making it E(8.6) min, and run the simulation model. Results for the average patient wait time are given in Fig. 2.2, compared for both steady-state and nonsteady-state cases.

It is clearly seen that the nonsteady-state wait time (upper curve) grows with no apparent trend to flattening out (plateau). The growth goes on indefinitely with time. In contrast, the steady-state wait time growth is limited, and it goes to the constant value plateau 25.3 min, which is identical to the value obtained from the queuing formulas.

A similar plot for the average number of patients in the queue for steady-state and nonsteady-state clinic operation is not shown here.

This example also illustrates an important principle of "unintended consequences." An intuition that is not supported by objective quantitative analysis says that a small change in the system input (service time from the average 8 min to 8.6 min) would result in a small change in the output (small increase in the number of waiting patients and their waiting time). For some systems, this is indeed true. Systems in which the output is always directly proportional to input are called linear systems.

However, there are quite a few systems in which this simple reasoning breaks down: a small change in the value of the system's input parameter(s) results in a dramatic change in the system's outcome (behavior), e.g., from a steady-state regime to a nonsteady-state regime with unlimited growth (sometimes, an opposite effect happens: a large change in the system input results in a very small and weak system output). Such systems are called nonlinear. They often exhibit complex behavior that is difficult to foresee without some type of numeric analysis despite the fact that such systems can consist of only a few elements.

2.3.3   Outpatient Clinic: Limited Queue Size with Leaving "Inpatient" Patients

Unlimited queue size is not always a good model of real systems. In many cases, patients wait in a waiting lounge that has usually a limited number of chairs (space). QA models designated *M/M/s/K* are available that include a limited queue size, *K* (Green 2006; Lawrence and Pasternak 2002; Hall 1990). However, analytic formulas become very cumbersome. If the QA model should also include some patients who leave the system after waiting for sometime in the queue before the service starts (typical situation for nonemergency patients), the analytic formulas become almost intractable (at least for those who do not have operations research background). Therefore, QA application is not presented here.

2.3.3.1   Discrete Event Simulation Application

In contrast to QA, DES models easily handle the limited queue size and patients leaving before the service starts ("inpatient" patients).

To illustrate, we use the same DES model as in Sect. 2.3.1.3 with only a slight modification to include a new limited queue size and "inpatient" patients.

Suppose that the queue size limit is 15 (the maximum number of chairs or beds), and patients leave (renege) after waiting from 10 to 25 min with the most likely 15 min (of course, these could be any other numbers or statistical distributions). We put 15 in the field "Input queue size," and draw a routing called "renege after *T*(10,

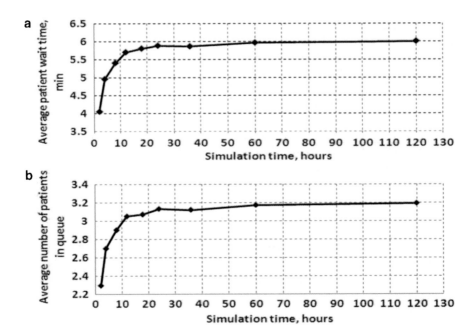

Fig. 2.3 Average patient wait time (**a**) and average number of patients in queue (**b**) for a clinic with limited queue size (15) and "inpatient" patients leaving the queue after waiting from 10 to 25 min with the most likely 15 min

15, 25) min." This represents a triangle distribution that is widely used in DES modeling as a good approximation when accurate data are not readily available. We keep the average service time 8.6 min that resulted in a nonsteady-state unlimited queue growth for the unlimited queue size (Sect. 2.3.1.3). The new model is now ready to go. Simulation results are presented in Fig. 2.3.

The difference between the nonsteady-state case with unlimited queue size (Sect. 2.3.2) and limited queue size with leaving "inpatient" patients is significant.

The plots suggest that limited queue size and leaving patients turn the nonsteady-state process into the steady-state (plateau). (It could be proved that a steady-state solution always exists if the queue size is limited). The steady-state average waiting time is about 6 min (top plot-a). The average steady-state number of patients in the queue is about 3.2 (bottom plot-b). This operational performance looks much better on the first glance than the original case with the unlimited queue size and no leaving patients.

However, the model's statistics summary also shows that about 10–11% of patients are lost because they left the queue after waiting too long. This means that a better operational performance in terms of waiting time and the number of patients in the queue is achieved only because fewer patients are served. Apparently, serving fewer patients is not typically an appropriate approach for improving operational performance. Besides, serving fewer patients results in lost revenue.

Thus, this simple DES model gives a lot of valuable information and serves as a powerful tool to analyze and predict the clinic operational performance (see also a more detailed example in Sect. 2.4.5 that includes an analysis of trade-off between clinic capacity, staffing, and net revenue).

2.3.4 Outpatient Clinic: Time-Varying Arrival Rates

In the previous Sect. 2.3.2, the queuing system with unlimited queue size (average arrival rate 28 patients per h and average shot time 8.6 min) was nonsteady-state one, for which QA model formulas could not be used.

However, the clinic's manager realized that the average patient arrival rate varies significantly during a day, and that 28 patients per h was actually a peak arrival rate, from noon to 3 p.m. In the morning hours from 8 to 10 a.m., the arrival rate was lower, 15 patients per h. From 10 a.m. to noon, it was 20 patients per h, and in the afternoon from 3 to 6 p.m. it was about 23 patients per h.

How is the time-varying arrival rate affect the clinic operational performance?

2.3.4.1 Traditional Approach

Because the arrival rate before noon is lower than the peak value in the midday and then patient arrival rate slows down again, this should somewhat compensate the peak value. The average arrival rate is going to be $(15+20+28+23)/4 = 21.5$ patients per h. Therefore, the clinic operational performance should be close to steady-state and acceptable.

2.3.4.2 Queuing Analytic Theory Application

In order to take into account the time-averaged arrival rate for these time periods for the day, the manager calculated a more accurate time-averaged arrival rate (rather than the simple arithmetic average) as $(15\times2+20\times2+28\times3+23\times3)/(2+2+3+3) = 22.3$ patients per h. He/she plugged this number in the queuing calculator (along with the average time to make a shot 8.6 min), and obtained for four servers the average number of patients in queue $L_q = 2.4$ patients, and the average waiting time about 6.4 min. Because the calculator produced some numbers, the manager concluded that the clinic process will indeed be in a steady-state condition, and that the waiting time and the number of patients in the queue is acceptable.

But is it a correct conclusion? Recall that QA models assume that a Poisson arrival rate is constant during a steady-state time period (Hall 1990; Lawrence and Pasternak 2002; Green 2006). If it is not constant, such as in this case, QA results could be very misleading. The wait time will be significantly greater in the midday period (and/or the steady-state condition will be violated). At the beginning and at the end of the day, though, the wait time will be much smaller. Because the arrival rate is included nonlinearly in the exponential term of a Poisson distribution

Table 2.1 QA summary results

Time period of the day	Arrival rate (patients per h)	Average number of patients in queue, L_q	Average wait time (min)
8–10 a.m.	15	0.25	1.0
10 a.m.–12 p.m.	20	1.15	3.45
12–3 p.m.	28	No steady-state solution	No steady-state solution
3–6 p.m.	23	3.0	7.8

formula, the arrival rate cannot be averaged first and then substituted in the exponential term. (For nonlinear functions, the average value of a function is not equal to the function of average values of its arguments).

As Green (2006) stated "…this illustrates a situation in which a steady-state queuing model is inappropriate for estimating the magnitude and timing of delays, and for which a simulation model will be far more accurate."

It is tempting, as a last resort, to save the use of QA model by dividing the day into time periods in which arrival rate is approximately constant. Then a series of *M/M/s* models is constructed, one for each period. This approach is called SIPP (stationary independent period-by-period) (Green 2004, 2006).

If we apply this approach, the following results presented in Table 2.1 can be obtained.

Notice how these results differ from those based on the averaging of the arrival rate for the entire day.

However, this SIPP patch applied to QA models was found to be unreliable (Green 2004, 2006). This is because, in many systems with time-varying arrival rates, the time of peak congestion significantly lags the time of the peak in the arrival rate. A modification has been developed called *Lag-SIPP* that incorporates an estimation of this lag. This approach has been shown to often be more effective than a simple SIPP (Green 2006).

Even it is so, this does not make QA models application less cumbersome if there are many time periods with different constant arrival rates because many different *M/M/s* models need to be constructed accordingly to describe one process.

It is illustrated below how DES model easily and elegantly handles this situation with time-varying arrival rate.

2.3.4.3 Discrete Event Simulation Application

The DES model structure (layout) for time-varying arrival rate is the same as it was used in Sect. 2.3.1. The only difference is a different arrival routing type: instead of periodic arrival with the random interarrival time, an input daily-pattern arrival panel should be used. We use 1 day of week, and input 30 patients from 8 to 10 a.m. (15 patients per $h \times 2$); 40 patients from 10 to noon (20 patients per $h \times 2$); 84 patients from noon to 3 p.m. (28 patients per $h \times 3$); and 69 patients from 3 to 6 p.m. (23 patients per $h \times 3$). The model of the entire day (from 8 a.m. to 6 p.m.) is ready to go.

Table 2.2 DES summary results

Time period of the day	Arrival rate (patients per h)	Average number of patients in queue, L_q	Average wait time (min)
8–10 a.m.	15	0.15	0.4
10 a.m.–12 p.m.	20	0.7	1.6
12–3 p.m.	28	4.6	8.8
3–6 p.m.	23	4.9	11.5

The simulation results are presented in Table 2.2 (compare with the approximated QA SIPP model results):

It is seen that the average number of patients in queue and the average wait time are accumulated toward the end of the day. Thus, QA SIPP model overestimates the queue at the beginning of the day and underestimates the queue at the end of the day. Of course, only DES model can provide results for the time period noon to 3 p.m.

2.3.5 "Excessive" ICU Capacity, "Improved" Efficiency, and Access to Care

There is a ten-bed ICU unit. The average daily patient arrival rate is 2 patients per day (but actual daily arrival rate varies from 1 to 3 patients depending on the day of the week). Patient length of stay (LOS) is in the range from 1 to 3 days, with 2.5 days being the most likely. It is observed that the daily average number of ICU patients (occupied beds) is five. The manager believes that the average daily utilization of five beds out of total available ten beds, i.e., 50%, is too low. The manager wants to improve the efficiency of the unit.

2.3.5.1 Traditional Management Approach

Because the available capacity of ten beds is not fully utilized, the manager decides to trim the "extra" capacity, i.e., to take out of service at least four "extra" beds leaving only six active beds (one bed above the daily needed average five is left as a precaution, just in case if it is suddenly needed).

This way, the daily utilization will be 5/6=83% instead of 50%. On top of that, because there is no need to staff four beds any more, the nursing and cleaning services budget will also be trimmed. This looks like a good management decision.

2.3.5.2 Queuing Analytic Approach

Due to variations around the average daily number of patients in the unit, this average value will be exceeded on a regular basis, and operational problems will occur regularly.

For example, a random sample with the average of five patients for 1 week can be:

Monday—7,   Tuesday—6,   Wednesday—4,   Thursday—8,   Friday—4, Saturday—3, and Sunday—3. Thus, there will not be enough capacity for 2 days of this particular week (Monday and Thursday), i.e., about 28% of time.

What is the probability that more than six beds will be needed with the average daily five patients, $P(\#beds > 6)$? Patient admissions are often (but not always) independent random time arrivals. A Poisson type process is widely used to model such random events. Using a Poisson formula, it is easy to find that

$$P(\#\text{beds} > 6) = 1 - \exp(-5)\sum_{n=0}^{n=6}\frac{5^n}{n!} = 24\%.$$

Thus, more than six beds will be needed for about a quarter of time. Therefore, the traditional management approach would create a regular bed shortage.

In order to apply QA formulas to estimate an average wait time and the number of waiting patients, a usual QA assumption should be made that the length of stay in the ICU is exponentially distributed and, of course, that patient arrival is a Poisson process.

Using $M/M/s$ model with the average interarrival rate $2/24 = 0.08333$ patients per h and the average service time (LOS) 2.5 days $\times 24 = 60$ h, it is easy to calculate that for the six bed ICU the average number of patients in queue is 2.9, the average waiting time is 35 h (about 1.5 days!), and the average ICU utilization is about 83%.

2.3.5.3   Management Engineering Approach

A more detailed and realistic analysis of unintended consequences of trimmed capacity can be performed using a very simple DES model without resorting to stringent assumptions needed for the use of analytic mathematical formulas of queuing theory. The DES model structure is similar to the one used in Sect. 2.3.1.

Let us demonstrate how different distributions with the same average for arrival and service time affect predicted operational performance. Recall that QA can be applied only for a Poisson arrival process and exponential service time (or approximately for distributions with the coefficient of variation close to 1; Green 2006). QA is severely limited in what it cannot account for different distributions of arrival pattern and service time with the same average; it always produces the same result if the same averages are used as input regardless of the effect of different distributions with the same average.

First, let us run the simulation using the same assumptions required for QA formulas: a Poisson arrival rate 2 patients per day and exponential LOS (service time) with the average 2.5 days (60 h). The steady-state result for the average number of patients in the queue is 3.1 and the 99% confidence interval for the average wait time is 28–34 h. This is very close to QA results. About 46% of patients wait more than 6 h. Recall that an exponential distribution is unbounded on high side; therefore, it can produce very large LOS values, such as 7, 8, or even 12 days LOS in this case. These large LOS values cause a large average wait time for new patients to get into ICU.

Second, let us run a more realistic simulation scenario with the same average LOS (service time) 2.5 days (60 h) but limited to the range from 1 to 3 days (fitted by a double-side-bounded triangle distribution). The steady-state simulation outcome for the average number of patients in the queue is 1 and the 99% confidence interval for the average wait time is 6.0–6.6 h. About 26% of patients wait more than 6 h. This is much better than predictions based on QA assumptions because LOS is now strictly limited to less than 3 days.

Now, instead of a Poisson arrival with the average rate 2 patients per day (interarrival time 0.5 day), let us use, for example, the following daily patient arrivals:

Monday—4,   Tuesday—4,   Wednesday—1,   Thursday—3,   Friday—1, Saturday—1, and Sunday—0.

Notice that the weekly average arrival rate remains the same as above: $(4+4+1+3+1+1+0)/7 = 2$ patients per day.

The simulation model layout is the same as in Sect. 2.3.1.3 except that the input daily arrival pattern is used instead of the periodic exponentially random arrival pattern. The steady-state simulation results in a weekly average number of patients in the queue about 0.6–1 and the average wait time 3.3 h. This is better than the above simulation scenarios and QA predictions but still is not good enough to meet safety and quality standards for many patients waiting admission into ICU.

This example illustrates a fundamental management engineering principle: because of variability of the number of patient arrivals and length of stay, some degree of reserved capacity (sometimes up to 40%) is needed in order to avoid regular operational problems (Green 2006).

Another possible way of improving the unit's operations is the daily load-leveling (smoothing) of the number of elective procedures scheduled for patients with required postsurgical ICU admission, as illustrated in Sect. 2.3.7 and also in Kolker (2009).

In summary, capacity and staffing decision/planning based only on averages without taking into account particular arrival and service time distributions usually result in significant miscalculations. This is called the flaw of averages (see also Costa et al. 2003; Marshall et al. 2005; de Bruin et al. 2007; Savage 2009).

2.3.6 Mixed Patient Arrival Patterns: Simultaneous Random and Scheduled Arrivals

Frequently mixed patient arrival patterns exist, i.e., some patients are scheduled to arrive at specific time while other patients arrive unexpectedly at random points in time. For example, some clinics accept patients who made an appointment, but also accept urgent random walk-in patients. Operating room suites schedule elective surgeries while suddenly a trauma patient arrives and an emergency surgery is required. Such a mixed patient arrival patterns with different degrees of the variability requires a special treatment.

Suppose that there is one procedure room and there are eight procedures scheduled for a day, at 6 a.m., 8 a.m., 10 a.m., 12 p.m., 2 p.m., 4 p.m., 6 p.m., and 8 p.m. On this day, six random emergency patients also arrived with the average interarrival time 4 h. Total average number of patients for 1 day is 14. The average procedure time is 1 h.

What is the operational performance of this simple unit in terms of the number of patients in the queue and their wait time?

2.3.6.1 Traditional Management Approach

If on average there are 14 patients per day, and an average procedure takes 1 h, then no patients in the queue and no wait time on that day (for 24 h) is expected at all.

2.3.6.2 Queuing Analytic Approach

QA models should not be used if arrival flow contains a nonrandom component, i.e., it is not a Poisson random arrival. Let us illustrate what happens if this principle is violated.

If $M/M/s$ QA model is applied assuming that all 14 patients are random arrivals, then the arrival rate is 14 patients/24 = 0.583 patients per h. Using the average procedure time 1 h, we get the average number of patients in the queue, $L_q = 0.82$, and wait time in queue, $W_q = 1.4$ h.

Notice that if traditional approach actually treats all values as fixed numbers with no variability, QA approach is another extreme: all values (even scheduled times) are considered random with exponential distribution (exponentially random).

2.3.6.3 Management Engineering Approach

Now, let us use a simple DES model with two arrival flows, one exponentially random, E(4) hours, and another one with scheduled eight patients, as indicated in Fig. 2.4.

The result of the steady-state two weeks DES simulation length is that the average number of patients in the queue was $L_q = 0.62$ and the average wait time in the queue $W_q = 1$ h.

Notice how QA model overestimates the average number of patients in the queue (almost by 32%) and the wait time (almost by 40%). This is a reflection of the general management engineering principle: the higher is the degree of randomness in arrival rate and service time, the lower is the unit operational performance in terms of the queue size, wait time, and utilization.

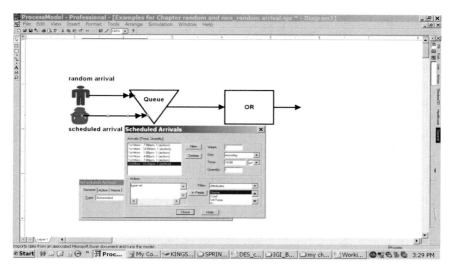

Fig. 2.4 Mixed patient arrival pattern: random and scheduled

Thus, it is again illustrated that QA models cannot account accurately enough for arrival variability that is lower than a Poisson variability, let alone a nonrandom scheduled arrivals.

A similar mixed patient arrival pattern will be used in next Sects. 2.3.7 and 2.3.8.

2.3.7   Small Rural Hospital vs. Large Community Hospital: Does Size Affect Operational Efficiency?

Hospitals or hospital departments often try to benchmark their operational performance against some other hospitals that are considered industry best practice. For example, if one hospital has low patient admissions wait time, then another hospital with higher admission wait often launches improvement project to meet that low admission target that another hospital already achieved. That hospital studies admission and other processes that help best achiever to get its impressive performance, and then is trying to implement what it learned hoping to get similar results.

Suppose that there are two hospitals: one is a small rural hospital with total 25 beds, and another one is a large community hospital with total 250 beds. The large hospital claims that it admits practically all patients with no waiting time. The small rural hospital acknowledges that its admission time is typically at least about 2 h, and that the significant percentage of patients waits more than 2 h. The small hospital wants to learn and implement the best admission practice from the large counterpart to get a similar low admission time.

2.3.7.1 Traditional Management Approach

During the benchmarking study it was observed that a typical patient length of stay for both hospitals was practically the same, from 1 to 8 days with the most likely 4 days.

Both hospitals had some scheduled patient admissions and some emergency arrivals.

It was observed that the small hospital typically scheduled one patient a day and had about four emergency admissions. The large community hospital typically admitted ten scheduled patients a day and had about 40 emergency admissions.

It turns out that the ratio of the total number of admitted patients to the number of hospital beds is exactly the same: 5 over 25 for the small hospital and 50 over 250 for the large hospital. The management of the small hospital reasoned that because those ratios are the same, the patient length of stay is also the same and the implemented admission process steps are similar for both hospitals their operational performance in terms of admission wait time should also have been the same, i.e., very low or not at all.

Yet, despite their best efforts, the waiting time for small rural hospital was still high enough, about 2 h. The management of the small hospital management was puzzled.

2.3.7.2 Management Engineering Approach

Because both hospitals have mixed patient arrival pattern (emergency random and scheduled) direct application of QA formulas should not be used (as it was demonstrated in the previous Sect. 2.3.6).

Instead, a simple DES model should be used with two arrival flows (exponentially random and scheduled arrivals) similar to the one indicated in Fig. 2.4.

DES model input for small hospital is capacity 25 beds, one patient arrival scheduled daily at fixed time and four exponentially random patient arrivals per day (average interarrival time is 6 h).

Model input for the large hospital is capacity 250 beds, ten patient arrivals scheduled daily at fixed time and 40 exponentially random patient arrivals per day (average interarrival time is 0.6 h).

The length of stay (LOS) for both hospitals is the same: minimal 1 day, the most likely 4 days and maximal 8 days, approximated by the triangle distribution.

The steady-state simulation results presented in Table 2.3 are instructive.

Large 250 bed hospital has indeed negligible admission wait time. Small 25 bed hospital has very large average admission wait time and the high percentage of patients waiting longer than 2 h (compared to the large hospital).

This is an illustration of an important fundamental management engineering principle: the performance of scale. Size does matter. Large hospitals (units) always have better operational performance characteristics (lower wait time and number of patients in the queue, higher utilization) than small units with the same input relative to its size. This is a reflection of a nonlinear scale effect that is typical for healthcare

Table 2.3 Comparative performance characteristics of two hospitals with different bed capacity and the same ratio of patient arrivals to their bed capacity (size)

	99% CI of average admission wait time (h)	99% CI of percentage of patients waiting longer than 2 h (%)	Average capacity utilization (%)
Large hospital, 250 beds	0.001–0.003	0.02–0.06	81
Small hospital, 25 beds	2.5–3.0	15.5–17.5	81
Small hospital, 25 beds, with reduced LOS	0–0.002	0.06–0.16	53

organizations. Failure to take this effect into account usually invalidates simple linear benchmarking and proportional adjustments.

If the small hospital wants to reduce its admission waiting time, but patient arrival rate is beyond its control and the number of beds is fixed, then the only option is reducing patient length of stay. For example, DES modeling results presented in the last line in Table 2.3 indicate that the admission wait time as well as the percent of patients waiting more than 2 h for the small 25 bed hospital will be practically negligible and similar to the large 250 bed hospital if the most likely length of stay (LOS) is cut down to 3 days and the maximal LOS does not exceed 4.5 days. However, in this case the average capacity utilization is also significantly reduced, which is usually an undesired effect. Lower utilization is the price paid for lower admission wait time. Attempts to increase utilization (desired effect) will inevitably result in increase of the admission wait time (undesired effect).

In summary, process improvement efforts based on linear benchmarking and simple linear proportional adjustments of input values could be misguided and short-lived if the scale effect (organization size) is not taken into account.

2.3.8 Daily Load-Leveling (Smoothing) of Scheduled Elective Procedures

In most hospitals, random (emergency) surgeries compete for the same operating rooms (OR) resources with scheduled (elective) surgeries. While the variable number of daily emergency surgeries is beyond hospital control (this is a natural variability), there is a significant variation in the number of daily scheduled elective surgical cases that could be actively managed using hospital scheduling system (Litvak and Long 2000; Kolker 2009).

It is possible to manage the scheduling of the elective cases in such a way that smoothes the overall patient flow variability. A daily load-leveling of elective cases would reduce the chances of excessive peak demand for the system's capacity (operating rooms and ICU) and, consequently, would reduce patient waiting time.

In 2011, The Leapfrog Group Hospital Survey included in the new section Patient Experience of Care the use of operational management methodology (management engineering) and, specifically, smoothing elective patient scheduling (Leapfrog 2011).

What is a quantitative effect of the daily load-leveling of elective surgeries on delay to start the case in the presence of the competing demand from random emergency surgeries for OR resources?

2.3.8.1   Traditional Management Approach

It is generally known that elective scheduling smoothing could help reduce delay (Litvak and Long 2000; McManus et al. 2003, 2004).

For example, McManus et al. (2003) concluded "…variability in scheduled surgical caseload represents a potentially reducible source of stress on the ICU in hospitals and throughput in the healthcare delivery system generally." Further, they stated "…the data demonstrated that bed availability was more strongly determined by variation in scheduled demand than by variation in requests for unscheduled admissions." The authors repeat several times that "…artificial variability is best managed by elimination wherever possible," and that "…we propose that hospitals first seek to control artificial variability as much as possible." However, the authors provide no specific information on how to quantitatively evaluate an effect of smoothing on availability and delay of care.

2.3.8.2   Queuing Analytic Approach

As it was demonstrated in the previous Sect. 2.3.5, QA cannot help in analyzing this problem because it cannot account for both nonrandom components and different distributions of arrival pattern and service time with the same average; it always produces the same result if the same averages are used as input regardless of the effect of different distributions with the same average.

2.3.8.3   Management Engineering Approach

While the number and timing of emergency cases are random by their nature, elective procedure scheduling is usually within the hospital management control. In order to quantitatively analyze the effect of daily load-leveling, it is required to make two simulation models: (1) baseline model that uses current elective and emergency admission schedules (mixed patient arrival pattern) to calculate the delay for emergency and scheduled patients; (2) model with load-leveled (smoothed) elective schedule and the same emergency admissions to calculate the delay for emergency and scheduled patients.

A comparison of the difference in the delay (if any) helps to make a conclusion.

An example of the number of elective and emergency admissions, as well as a possible smoothed (load-leveled) elective schedule for 4 weeks time period, is given in Table 2.4.

A smoothed schedule (last column in Table 2.4) has the same number of cases over the 4-week period as the original unsmoothed schedule, i.e., not a single case

Table 2.4 Elective, emergency, and daily load-leveled admissions for the 4-week period

Week	Day of week	Number of elective admissions	Number of emergency admissions	Number of daily-leveled (smoothed) elective admissions
1	Monday	9	16	6
1	Tuesday	11	14	7
1	Wednesday	8	14	7
1	Thursday	5	20	7
1	Friday	5	15	7
2	Monday	10	18	7
2	Tuesday	13	20	7
2	Wednesday	11	9	7
2	Thursday	8	11	7
2	Friday	3	20	7
3	Monday	5	17	7
3	Tuesday	9	11	7
3	Wednesday	8	15	7
3	Thursday	6	15	7
3	Friday	6	20	7
4	Monday	7	15	7
4	Tuesday	4	13	7
4	Wednesday	3	12	7
4	Thursday	4	11	7
4	Friday	3	20	6
	Total	138	306	138

was dropped but they are rather rearranged over the time period to smooth daily peaks and valleys.

It was assumed that three interchangeable operating rooms (ORs) are available in this case. The emergency surgery duration is in the range from 1.5 to 2.5 h, with this most likely time of 2.1 h. Elective surgeries duration is in the range from 1.5 to 3 h, with the most likely time of 2.4 h. Both are represented by corresponding triangle distributions similar to those used in the previous sections.

Emergency patients were assumed to arrive uniformly randomly 24 h during a day. Elective surgeries were scheduled daily from 7 a.m. to 3 p.m.

Simulation model layout is similar to the one presented in the previous section (Fig. 2.4).

Simulation for 672 h (4 weeks) for the original unsmoothed elective schedule along with competing emergency cases resulted in the average patient waiting time of 0.74 h for emergency cases (99% CI is 0.73–0.76 h) and 1.2 h for elective cases (99% CI is 1.19–1.24 h).

The same simulation modeling with a smoothed (load-leveled) elective schedule along with the same competing emergency cases resulted in the average patient waiting time of 0.58 h for emergency (99% CI is 0.57–0.6 h) and 0.82 h for elective cases (99% CI is 0.8–0.84 h).

Thus, in this particular example, elective daily load-leveling results in about 21% reduction in waiting time for emergency surgeries and about 32% reduction in waiting time for elective cases.

Elective schedule smoothing (daily load-leveling) is indeed a very powerful approach of reducing patient waiting time and improving efficiency. This is a fundamental management engineering principle.

A simple simulation model also allows testing an effect of the different smoothing schemes. For example, if nearly the same daily number of elective cases is not possible due to some practical limitations, it is possible to test another less perfect smoothing scheme to make sure that the end result is still worth the effort of its implementation (or maybe not). No traditional management methods are capable of providing such insights for decision-making.

Kolker (2009) provided a detailed analysis of the effect of daily load-leveling of elective surgeries on ICU performance. The author addressed the following question: "What maximum number of elective surgeries per day should be scheduled along with the competing demand from emergency surgeries in order to reduce diversion of an ICU with fixed bed capacity?" The simulation model led to the conclusion that a "cap" of 5 cases per day was needed with extra overflow cases scheduled within two weeks using unfilled block time.

Ryckman et al. (2009) reported the results of practical implementation of load-leveling of elective surgical admissions at Cincinnati Children's Hospital. New elective surgical admissions to the pediatric ICU were also restricted to a maximum of 5 cases per day. As a consequence of the smoothing of elective surgical cases, it was observed that there was a near elimination of ICU diversion and cancelation of elective surgeries due to lack of ICU beds.

2.3.9   Separate or Interchangeable (Shared) Operating Rooms for Emergency and Scheduled Surgeries: Which Arrangement Is More Efficient?

The issue of specialized vs. pooled (interchangeable) Operating Rooms (ORs) caused a controversy in literature on healthcare improvement. Specifically, if surgical cases include both scheduled (elective) and unscheduled emergency surgeries starting at random time, is it more efficient to reserve specialized OR dedicated separately for scheduled and emergency surgeries, or is it better to perform both types of surgeries in any available OR using pooled or interchangeable OR arrangement?

2.3.9.1   Traditional Management

Haraden et al. (2003) recommend that hospitals that want to improve patient flow should designate separate ORs for scheduled and unscheduled (emergency) surgeries. The authors state that in this arrangement "...Since the vast majority of surgeries is scheduled, most of the OR space should be so assigned. Utilization of the scheduled rooms becomes predictable, and wait times for unscheduled surgery become manageable." The authors imply that this statement is self-evident, and provide no quantitative analysis or any justification for this recommendation.

2.3.9.2 Queuing Analytic Approach

QA cannot help in addressing this problem because of its limitations discussed in the previous sections.

2.3.9.3 Management Engineering Approach

In contrast to the above-described traditional approach, Wullink et al. (2007) developed a DES model of OR suite for the large Erasmus Medical Center hospital (Rotterdam, The Netherlands) to quantitatively test scenarios of using specialized dedicated ORs for emergency and for scheduled surgeries vs. pooled (interchangeable) ORs for both types of surgeries. These authors concluded that based on DES model results "…Emergency patients are operated upon more efficiently on elective ORs instead of a dedicated emergency ORs. The results of this study led to closing of the emergency OR in this hospital."

In contrast to an intuitive "common sense" recommendation of Haraden et al. (2003), Wullink et al. (2007) presented specific data analysis to support their conclusions: pooled use of all ORs for both types of surgery results in the reduction of the average waiting time for emergency surgery from 74 to 8 min.

In this section, a simple generic simulation model is presented to address the same issue and to verify literature results. For simplicity, we consider an OR suite with two operating rooms, OR1 and OR2. Both emergency (random) and scheduled surgeries are included.

Let us first consider the situation when the majority of cases are scheduled elective surgeries. Six elective surgeries are scheduled 5 days a week, Monday to Friday at 7 a.m., 9 a.m., 11 a.m., 1 p.m., 3 p.m., and 5 p.m.

Emergency surgeries are assumed to start independently randomly 24 h a day with the average patient interarrival time of 6 h (a Poisson arrival rate is 0.166 patients per h).

Similarly to the values used in the previous section, the most likely scheduled surgery duration is assumed to be 2.4 h, and emergency surgery duration is assumed to be 2.1 h (Wullink et al. 2007). The assumed variability is from 1.5 to 3 h for scheduled surgeries and from 1.5 to 2.5 h for emergency surgeries.

Using these arrival and service time data, let us consider two scenarios.

Scenario 1: there are two ORs, one is specialized dedicated only for scheduled surgeries (OR1) and stays open from 7 a.m. to 7 p.m.; another one is specialized dedicated only for emergency surgeries (OR2) and is open 24 h a day, as shown in

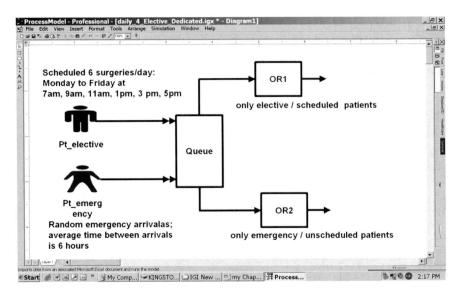

Fig. 2.5 Simulation model layout of dedicated operating room for scheduled elective surgeries OR1 and dedicated operating room for emergency unscheduled surgeries OR2

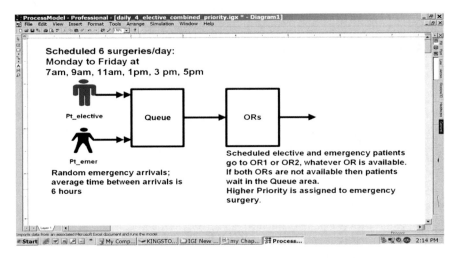

Fig. 2.6 Simulation model layout of interchangeable emergency and scheduled elective operating rooms (ORs)

Fig. 2.5. If the dedicated OR is not available, then the new patient waits in the queue area until the corresponding dedicated OR becomes available.

Scenario 2: there are also two ORs. However, they are pooled, i.e., fully interchangeable: both emergency and scheduled patients go to any available OR, as indicated in Fig. 2.6. If both ORs are not available, then patients wait in the queue area until one

Table 2.5 Specialized ORs vs. pooled ORs

Characteristics	Dedicated OR		Pooled OR	
	Elective	Emergency	Elective	Emergency
95% CI of the number of performed surgeries	28–29	20–21	30	20–21
95% CI of the average wait time, hours	2.3–2.9	0.4–0.49	0.14–0.17	0.16–0.2
95% CI of the average number of patients in Queue	0.65–0.8	0.08–0.1	0.035–0.044	0.03–0.04
Average weekly OR utilization (%)	99%	35%	46%	

Most surgeries are scheduled elective

of the ORs becomes available. Emergency patients have higher priority and move first to the available OR.

Simulation was performed for 5 days (120 h) Monday to Friday using 300 replications. Results for these two scenarios are given in Table 2.5.

An examination of the results is instructive. While the number of performed surgeries is practically the same for both scenarios, the wait time for scheduled surgery for pooled ORs is more than an order of magnitude (!) lower than that for dedicated ORs. The wait time for emergency surgery is also lower for pooled ORs by about a factor of 2. Practically no patients wait in queue for pooled ORs. This means that, contrary to expectation of Haraden et al. (2003), specialized dedicated ORs arrangement results in inevitable overtime in order to perform the same number of surgeries as in the pooled (interchangeable) ORs arrangement.

The reason for such an unexpected result for OR specifically dedicated to elective scheduled surgeries is the actual surgery duration variability. Therefore, scheduling elective surgeries every 2 h inevitably results in delays and/or required OR overtime. At the same time scheduled surgeries cannot be moved into another standby OR that is dedicated only for emergency patients, even though this OR is currently available. This situation is also reflected by the highly uneven weekly OR utilization, 99% and 35%, respectively.

Now, let us consider the situation when the majority of cases are unscheduled emergency surgeries. We have the same two scenarios with two ORs (dedicated and pooled) with the same surgeries duration variability.

However, this time only four daily elective surgeries are scheduled 3 days a week, Tuesday, Wednesday, and Thursday at 8 a.m., 10 a.m., 1 p.m., and 3 p.m. Dedicated ORs for scheduled surgeries stay open from 8 a.m. to 6 p.m. Emergency (random) surgeries are more frequent, with the average patient interarrival time of 2 h, Monday to Friday, 24 h a day.

Simulation results for 120 h (5 days) using 300 replications are given in Table 2.6.

Notice that in this case the number of performed emergency surgeries is higher for pooled ORs. Although the average waiting time for scheduled elective surgeries increased up to about 1 h for pooled ORs, the average wait time for emergencies dropped dramatically by more than an order of magnitude, from about 7 h down to

Table 2.6 Specialized ORs vs. pooled ORs

Characteristics	Dedicated OR		Pooled OR	
	Elective	Emergency	Elective	Emergency
95% CI of the number of performed surgeries	12	53–54	12	59–60
95% CI of the average wait time (h)	0.16–0.18	6.3–7.7	0.96–1.3	0.55–0.65
95% CI of the average number of patients in Queue	0.017–0.02	3.5–4.4	0.11–0.14	0.3–0.35
Average weekly OR utilization (%)	55%	92%	62%	

Most surgeries are unscheduled emergency

about 0.5 h. (Compare this dramatic drop to Wullink et al. (2007) result.) A similar picture is observed for the average number of patients in the queue.

Overall, these results support the conclusions of Wullink et al. (2007) that performing emergency surgeries in the interchangeable OR scenario is more efficient than in the reserved dedicated emergency OR. These authors provided a detailed instructive discussion on why the dedicated OR scenario performs worse, especially for emergency surgeries, while intuitively it seems that it should perform better, as Haraden et al. (2003) assumed.

Wullink et al. (2007) pointed out that besides reserving OR capacity for emergency surgeries arrival, ORs need to reserve capacity to cope with the variability of surgery duration. In the pooled ORs scenario, the reservation might be shared to increase the flexibility for dealing with unexpected long case duration and emergency surgery, whereas the dedicated scenario does not offer the opportunity to use the overflow principle.

On top of that, a dedicated OR scenario may cause queuing of emergency surgeries themselves because of their random arrival time. If emergency surgeries were allocated to all available ORs (pooled ORs scenario), then it would be possible to perform them simultaneously, thereby reducing the waiting time.

Wullink et al. (2007) acknowledge that "...interrupting the execution of the elective surgical case schedule for emergency patients may delay elective cases. However, inpatients are typically admitted to a ward before they are brought to the OR. Although delay due to emergency arrivals may cause inconvenience for patients, it does not disturb processes in the OR."

As simulation modeling indicates, delay in scheduled cases in the pooled ORs (if the majority of surgeries are emergencies) is usually not too dramatic (e.g., up to 1 h), while reduction of waiting time for emergency surgeries is very substantial (from 74 to 8 min according to Wullink's model, or from about 7 to 1 h, according to our simplified simulation model with generic input data described in this section).

If the majority of surgeries are scheduled, then there is not much delay in pooled ORs at all, both for scheduled and emergency surgeries.

At the same time, it is still possible a situation when the pooling of resources is not always beneficial with regard to the waiting times for urgent patients. This can happen if there is a relatively large difference in process (surgery) time for different types of patients, or due to significantly different performance targets (waiting time) because of a different level of urgency. A trade-off line can be developed that indicates whether the use of pooled or separate resources is more beneficial. For example, Joustra et al. (2010) demonstrated that the separation of urgent and regular patients in a radiotherapy outpatient department becomes beneficial if the target wait time for urgent patients is much shorter than that for regular patients.

2.3.10   Surgical Capacity of Special Procedure Operating Rooms

A specialized procedure operating room (SPR) unit is being planned to unload the volume of outpatient day surgeries from the main surgical department's general operating rooms. At the end of the current year, 1,844 special surgical procedures of type 1 and type 2 were performed. Projected additional procedure volumes for the next year is 179 procedures of type 1 and 13 procedures of type 2.

Patients get prepared for procedures in the preparation bed area; then they move to the available SPR for the procedures and come back into the same bed area (not necessarily into the same bed) for postprocedure recovery. After initial postanesthesia recovery time (phase 1), inpatients are moved to a regular nursing unit for full recovery. Outpatients stay in the bed area for the full recovery time (phase 1 and phase 2).

The average postprocedure recovery time is 3.3 h. The average time to perform a procedure is 0.82 h. SPR turnover time ranges from 10 to 20 min (cleaning the room, restocking it with supplies, and making it ready for the next patient). Bed turnover time (cleaning, changing linens, and making the bed ready for the next patient) is also from 10 to 20 min. The SPR unit is supposed to work annually 255 days from 7 a.m. to 5 p.m. (10 h daily, no weekends); the average annual utilization target is 85%.

The following operational SPR performance criteria have been established: (1) patient wait time to get into SPR for a procedure is not more than 1 h for 95% patients; (2) postprocedure patient wait time to get back into the recovery bed from SPR is not more than 5 min for 95% patients.

Management should decide on the minimal number of beds and SPR that are needed in order to meet the established operational performance criteria. This information is required by an architectural firm hired to design a floor plan needed for the unit construction.

2.3.10.1   Traditional Management

The number of beds and the number of SPR is calculated using a simple formula:

$$\# \text{beds} = \frac{\text{Annual\_Patient\_Volume} \times \text{Average\_Bed\_Time}}{\text{Total\_Annual\_Available\_Time} \times \text{Utilization}}$$
$$= \frac{2,036 \times (3.3 + 0.25)}{2,550 \times 0.85} = 3.33 \cong 4,$$

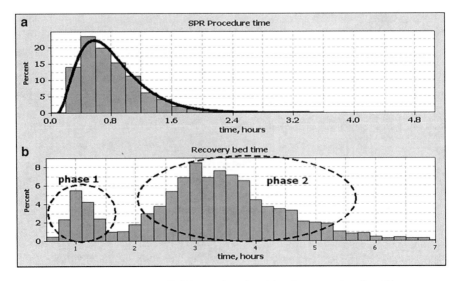

Fig. 2.7 Special procedure room (SPR) procedure time (**a**) and recovery bed time (**b**)

$$\# SPR = \frac{Annual\_Patient\_Volume \times Average\_(SPR + TurnOver)\_Time}{Total\_Annual\_Available\_Time \times Utilization}$$

$$= \frac{2,036 \times (0.82 + 0.25)}{2,550 \times 0.85} \cong 1.$$

Thus, four preparation and postprocedure recovery beds and one procedure room would be enough.

Because the projected annual patient volume will be served for the available annual time, no or very little patient wait time is expected.

2.3.10.2   Management Engineering Approach

The typical SPR procedure time and patient recovery bed time have wide and skewed distributions with long tails that are much larger than the average values, as indicated in Fig. 2.7 (panels a and b). Therefore, the use of the average time in the above formulas produces a misleading estimation of the required number of beds and SPR (resources).

The only way of capturing this wide time variability is by using an operational simulation model. The model layout is presented in Fig. 2.8.

The model incorporates patient preparation time variability (best fit statistical distributions) separately for inpatients and outpatients. The postprocedure recovery time is also incorporated separately for inpatients and outpatients, as well as the SPR best fit procedure time variability. Simulation modeling results are presented in Table 2.7.

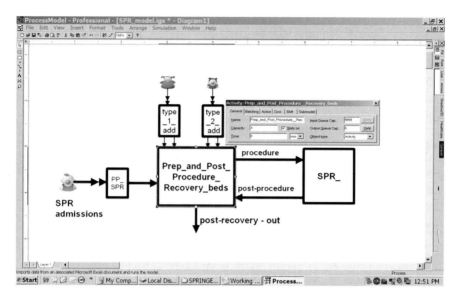

Fig. 2.8 Simulation model layout of patient flow for special procedure operating rooms (SPRs) and preparation and postprocedure recovery beds

It follows from this table that four beds and one special procedure room calculated using the average preparation and recovery and procedure time are badly underestimated: 23% of patients will wait longer than 1 h to get to SPR, and 23% of patients will wait longer than 5 min to get back into bed from SPR for recovery (vs. acceptable limit 5%).

A simulation model that takes into account random variability indicates that the minimal six beds and two SPR are required to meet operational performance criteria. In this case only about 2% of patients will wait longer than the acceptable limit.

Thus, the correct amount of resources in healthcare settings with highly variable demand can only be predicted by using process simulation methodology. This is another illustration of the flaw of averages, as already discussed earlier.

2.3.11 The Entire Hospital System Patient Flow: Effect of Interdependency of ED, ICU, OR, and Regular Nursing Units on System Throughput

A typical large community hospital includes main units: an Emergency Department (ED) with 25 beds; an intensive care unit (ICU) with 49 beds; a surgical department with 12 operating rooms (OR); and regular nursing units (NU) with a total capacity of 360 beds.

Table 2.7 The effect of the number of preparation and postprocedure recovery beds and special procedure rooms (SPR) on operational performance characteristics

Number of beds	Number of SPR	Average patient wait time to get to SPR (min)	Percent of patients waiting longer than 1 h to get to SPR (%)	Average postprocedure wait time to get back to bed for recovery (min)	Percent of patients waiting longer than 5 min to get back to bed for recovery (%)	Performance criteria met?
4	1	53	23	9.5	23	No
6	2	9.2	2.2	5.5	2.2	Yes

Total typical monthly patient volume is about 4,478 that include ED patients transported by ambulance, ED walk-in patients, and surgical scheduled patients.

The overall hospital performance needs significant improvement. A large percent of time ED is on ambulance diversion and there is a long ED patient line and long wait time. The ICU frequently does not have beds for ED patient admissions or delays admission of postsurgical patients. The surgical department is often at capacity, and elective surgeries are frequently canceled or rescheduled.

The hospital management needs to decide: what unit/department to start with for process improvement projects; what type of projects to select; and decide on process improvement performance metrics.

2.3.11.1 Traditional Management

Because the most patient crowding is visible in ED, it is believed that inadequate ED capacity is an issue. The management wants to increase ED throughput and capacity by reducing ED patient length of stay (ED LOS). A process improvement team is formed. After a lot of invested time and effort, the project improvement team finally reports a significant reduction of the average length of stay (ALOS) and reduction of ED diversion time. The management praises a great ED improvement success.

However, both the OR and ICU start reporting increased patient wait time to get in due to "no available OR or no ICU beds." This, in turn, results in increasing the cancelation rate for scheduled surgeries and keeps more postsurgical patients boarded in OR waiting for ICU beds. Hence, more surgical cases are now delayed.

The hospital management wants to repeat an ED improvement success by initiating ICU or surgical department process improvement projects. But the management is not sure anymore that departmental process improvement will be translated into the overall hospital patient flow improvement. They are looking for a better analysis and a sustainable hospital-wide solution.

2.3.11.2 Management Engineering Approach

The entire hospital system consists of interdependent departments/units that interact with each other. ED is not a stand-alone unit. Increased patient volume coming out of ED not always can consistently be supported by available OR and ICU capacity to handle it. Therefore, ED improvement is not necessarily translated into the overall hospital system improvement, although the overall hospital performance improvement is the actual goal. It turns out that patient flow is a property of the entire hospital system rather than the property of the separate departments/units. A detailed analysis is required of the overall hospital system patient flow and the interdependency of subsystems/units in order to establish the system weak link and

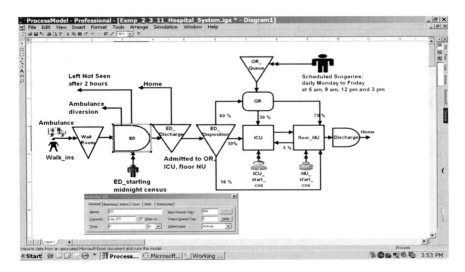

Fig. 2.9 Simulation model layout of patient flow for a typical entire hospital system

the right units for process improvement projects. Such an analysis can only be performed using a system simulation methodology (Kolker 2010, 2011).

Large systems are typically deconstructed into smaller subsystems using natural breaks in the system: emergency, surgical, ICU, floor nursing units, etc. The subsystems can be modeled and analyzed separately. However, they should be reconnected back in a way that recaptures the most important interdependency between them. Analysis of a complex system is usually incomplete and can be misleading without taking into account subsystems' interdependency.

A high-level flow map (layout) of the entire hospital system is shown in Fig. 2.9.

Patients transported into the ED by ambulance (~18%) or walk-in patients (~82%) form an ED input flow. Some patients are treated, stabilized, and released home. ED patients admitted into the hospital (ED output) form an inpatient input flow into the ICU, OR, and/or NU. The length of stay distribution best fit was identified separately for patients released home and patients admitted to the hospital (Kolker 2008). Patients waiting longer than 2 h in the ED waiting room leave the ED without being seen (LNS: lost-not-seen patients). About 60% of admitted patients are taken into operating rooms (OR) for emergency surgery, about 30% of admitted patients move into the ICU, and about 10% of patients are admitted from ED into the floor nursing units (NU).

When ED, OR, ICU, or NU are completely full (at full capacity), a diversion status is declared. The units stay on diversion until at least one bed in the unit is freed. Total unit diversion is defined here as the percentage of operational time when the unit is at full capacity and can no longer accept new patients.

OR suite has 12 interchangeable operating rooms used both for ED and scheduled surgeries. There are four daily scheduled OR cases at 6 a.m., 9 a.m., 12 p.m., and 3 p.m., Monday to Friday (there are no scheduled surgeries on weekends). Scheduled cases form a separate OR admission flow, as indicated in the diagram (Fig. 2.9).

Elective surgery duration depends on surgical service type, such as general surgery, orthopedics, neurosurgery, etc. For simplicity of this particular model, elective surgery duration was weighted by each service percentage, and the best statistical distribution fit was identified (inverse Gaussian in this case). Emergency surgery duration was best fit by Pearson 6 statistical distribution.

About 30% of postsurgery patients are admitted from OR into the ICU (direct ICU admission), while 70% are admitted into the floor NU. However, some patients (about 5%) are readmitted from the floor NU back to the ICU (indirect ICU admission from OR). ICU length of stay is assumed to range from 1 to 3 days, with the most likely of 1.5 days, represented by a triangle distribution. Kolker (2009) developed a detailed ICU simulation model and analysis.

Patient length of stay (LOS) in NU is assumed to range from 2 to 10 days, with the most likely of 5 days, also represented by a triangle distribution. At the simulation start, the ED, ICU, and NU were prefilled with the midnight census of 15, 46, and 350 patients, respectively (Prefilling is used instead of a simulation warm-up period in this example in order to approximate the steady-state regime).

A summary of simulation results for the hospital system is given in Table 2.8. Simulation was run for 31 full days (744 hours) using 300 replications.

There are nine performance metrics (99% Confidence Intervals—CI) indicated in column 1.

Baseline metrics that correspond to patient ED LOS up to 24 h are presented in column 2. Aggressive improvement efforts in the ED resulted in reducing LOS for patients admitted into the hospital to less than 6 h (from ED registration to ED discharge). However, because of interdependency of the downstream units, four out of nine metrics became worse (column 4). The ED bottleneck just moved downstream into the OR and ICU because of their inability to handle the increased patient volume from ED.

Thus, aggressive process improvement in one subsystem (ED) resulted in a worsening situation in other interrelated subsystems (OR and ICU). If, instead of too aggressive ED LOS reduction, a less aggressive improvement is implemented, e.g., ED LOS not more than 11 h for patients admitted to the hospital, then none of the nine metrics become much worse than the baseline state (columns 5 and 6). While in this case ED performance is not as good as it could be, it is still better than it is at the baseline level. At the same time, a less aggressive local ED improvement does not, at least, make the ICU, OR, and floor_ NU much worse. In other words, the less aggressive ED improvement is more aligned with the ability of the downstream units to handle the increased patient volume.

Thus, from the entire hospital system standpoint, the primary focus of process improvement should be on the ICU because of its highest percent of patients waiting admission more than 1 h followed by the OR and ED. At the same time, ED target patient LOS reduction program should not be too aggressive. It should be closely coordinated with that for OR and ICU. Otherwise, even if the ED makes a

Table 2.8 Summary of simulation results for the hospital system patient flow performance metrics

1	2	3	4	5	6
Performance metrics	Baseline state	Too aggressive ED improvement: patient LOS within 6 h	Better or worse than baseline?	Less aggressive ED improvement: patients LOS within 11 h	Better or worse than baseline?
99% CI of the average patient wait time to get to ED (min)	42–43 min	5.4–6 min	Much better	38–39 min	Better but not much
99% CI for ED diversion	23.9–24.1%	5.9–6.2%	Much better	22.6–22.8%	Better but not much
99% CI of the percentage of patients left not seen (LNS) after waiting in ED more than 2 h	7.0–7.4%	0.03–0.06%	Much better	5.6–5.8%	Better but not much
99% CI of the percentage of patients waiting admission to OR from ED longer than 1 h	44–46%	53–55%	*Worse*	46–48%	Not much different
99% CI for OR diversion	16.8–17.7%	21–22%	*Worse*	17.5–18.3%	Not much different
99% CI of the percentage of patients waiting admission to ICU from ED longer than 1 h	62–64%	71–73%	*Worse*	63–66%	Not much different
99% CI for ICU diversion	24.7–25.8%	30–31%	*Worse*	25.4–26.5%	Not much different
99% CI of the percentage of patients waiting admission to floor_NU from ED longer than 1 h	35–37%	35–37%	No difference	34–37%	Not much different
99% CI for floor_NU diversion	10.0–10.2%	10.1–10.4%	Not much different	10.1–10.4%	Not much different

significant progress in its patient LOS reduction program, this progress will not translate into improvement of the overall hospital-wide patient flow. Of course, many other scenarios could be analyzed using the simulation model to find out how to improve the entire hospital-wide patient flow rather than that for each separate local subsystem/unit.

Patient flow (throughput) is a general dynamic supply and demand balance problem. This is not a one-time snapshot. The system's behavior depends on time. There are three basic components that should be accounted for in these type of problems: (1) the number of patients (or, generally, any items) entering the system at any point of time (admissions); (2) the number of patients (any items) leaving the system at any point of time after spending some variable time in the system (discharges), and (3) limited capacity of the system which limits the flow of patients (items) through the system. All three components affect the flow of patients that the system can handle. A lack of the proper balance between these components results in the system overflow, bottlenecks or, sometimes, underutilization. Simulation methodology provides the only means of quantitative analysis of the proper balance and dynamic variability in complex systems.

This analysis illustrates the following fundamental management engineering principles: (1) improvement of the separate subsystems (local optimization or local improvement) does not necessarily result in the improvement of the entire system, (2) a system of local improvements (local optimums) could be a very inefficient system (Goldratt and Cox 2004), and (3) analysis of an entire complex system is usually incomplete and can be misleading without taking into account the *subsystems' interdependency*.

2.4   Scheduling and Staffing Problems

2.4.1   Scheduling Order for Appointments with Different Duration Variability

An outpatient clinic manager has to schedule two groups of patient appointments: one group is new patient appointments and another group is follow-up patient appointments. On average, a new patient appointment takes about 60 min but could be in the range from 45 to 90 min. The follow-up appointment takes on average 30 min and its duration is usually from 25 to 35 min.

On a typical day, five new and six follow-up appointments are scheduled. Using the average appointment time 60 and 30 min accordingly, it is estimated that the clinic's total work time will be $5 \times 60 + 6 \times 30 = 480$ min, or 8 h.

The following managerial problem arises: does appointment order affect patient wait time and clinic daily total work time? In other words, is it better to schedule new appointments first and then the follow-up appointments, or the way around? Or does the appointment order make a difference at all?

Table 2.9 Simulation of scheduling rules for one clinic operational day

Scheduling rule	99% CI of the average patient wait time (min)	99% CI of the number of patients with NO wait	99% CI of the total clinic time (h)
Smallest variability first	6.0–7.3	3.1–3.6	9.6–9.9
Random order	9.4–11.6	2.6–3.1	10.2–10.8
Largest variability first	17.7–21.6	1.6–2.0	11.9–12.8

2.4.1.1   Traditional Management Approach

All other factors being equal, the schedule slots should be filled in the order in which they are received (whoever first called for an appointment gets the first available slot).

For example, if a new patient is scheduled first at 8 a.m. and appointment time is 60 min on average, then the next follow-up patient will be scheduled at 9 a.m., and next patient will be scheduled at 9:30 a.m., and so on. If the follow-up patient is scheduled first at 8 a.m. and appointment time is 30 min on average, then the next new patient will be scheduled at 8:30 a.m., and so on. There is no preference for appointment order. Indeed, the total clinic time for the average 60 min appointment followed by 30 min appointment would be the same as that for a 30 min appointment followed by a 60 min appointment.

2.4.1.2   Management Engineering Approach

The appointment duration times of 60 and 30 min are only the average values. There is a significant variability around these averages that affect the clinic's operational performance. In order to capture an effect of the variability, a model of clinic operations should be developed using simulation methodology. The outcomes of scheduling scenarios can be quantified, and comparative conclusions on their effectiveness can be made.

The simulation model layout is very simple and straightforward for this particular case (see Sect. 2.3.1.3). Every patient (entity) has a descriptive attribute "appointment type": new or follow-up. If the attribute is "new," then the appointment variability is captured using a simple triangle statistical distribution from 45 min with the most likely value being 60 min to the maximum 90 min; if the attribute is "follow-up" then the appointment variability is captured using a triangle distribution from 25 min with the most likely value being 30 min to the maximum 35 min. A total of 11 appointments are included: five new and six follow-up appointments. For simplicity, simulation for one typical operational day is performed. Results are presented in Table 2.9.

Thus, appointment order does make a difference. Scheduling rule with the smallest variability first is much better both in terms of lower patient wait time and a higher number of patients who do not wait at all. This is in contrast to traditional management expectation of no effect of appointment order.

Table 2.10 Summary of 25 days of clinic operations

Scheduling rule	Average wait time (min)	Standard deviation of wait time (min)	Average number of patients with no wait	Standard deviation of number of patients with no wait
Smallest variability first	6.3	4.4	3.2	2
Random order	12.2	12.1	2.5	1.9
Largest variability first	19.2	17.4	1.4	1.6

Notice also that the total clinic work time significantly exceeds 8 h expected based on the average appointment duration, i.e., overtime is required to serve all scheduled patients (unless there is a nonfilled cancelation or no-show). This is a consequence of the appointment duration variability around the average, not the average duration itself.

Another example of the effect of appointment sequence is presented by the Institute for Healthcare Improvement on its website (IHI 2011). This example is based on Monte-Carlo simulation for 25 days of clinic operations. It is assumed that new appointments take 45 min, plus or minus 15 min (30–60 min range); follow-up appointments take 30 min, plus or minus 5 min (25–35 min range). Results for one random sample of 25 days are represented in Table 2.10.

These data also demonstrate that appointment sequence with the smallest variability first (follow-up appointments) results in the smallest patient wait time and the largest number of patients who do not wait at all, followed up by appointment sequence with random order and largest variability first (new appointments). Similar conclusion has been made by Klassen and Rohleder (1996), Cayirli et al. (2006), Teow (2009), and Ben-Arieh and Wu (2011).

These examples illustrate a fundamental management engineering principle (practically proven in manufacturing): scheduling appointments (jobs) in the order of increased variability (jobs with lower variability come first) results in a lower overall cycle time and patient wait time.

The reason for the erroneous traditional management decision-making is, already mentioned, lacking a means to take into account the effect of the appointment length variability around the average.

2.4.2   Centralized Discharge vs. Individual Units Discharges

Inability to discharge patients in a timely fashion is a typical hospital problem.

Let us consider an ICU with four nursing units. Discharges occur daily Monday to Friday, usually in the afternoon between 2 and 9 p.m. The number of discharges from each unit is a random quantity with the average four daily discharges, i.e., the number of discharges on a particular day could be less or more than four. The most likely time to complete discharge is about 30 min, but the time could range from

20 min to as high as 60 min. So, on average about 2 h of nursing time is spent on the discharge process for each of four units. This nursing time is taken away from direct patient care. The ICU manager wants nurses to spend more time for direct patient care rather than for discharge paperwork.

2.4.2.1 Traditional Management Approach

After a brainstorming session, the idea of hiring a dedicated discharge nurse was proposed. This was considered a good management solution for all units because (PHLO 2008):

1. The unit nurse would be free from a paper-intensive discharge process to maximize their time for direct patient care.
2. Bed management could rely on one contact person who would notify cleaning services to clean beds after discharge.
3. Case management would get a resource to help coordinate and plan their activities.
4. Patient flow and throughput would improve because of a more timely discharge process.

2.4.2.2 Management Engineering Approach

The centralized discharge should be analyzed quantitatively and then compared to the performance of the current process before creating an additional discharge nurse position. Let us consider two scenarios.

Scenario 1. Current process: there are four independent units. Each unit has its own nurse who handles discharges as they ordered. Four random discharges take place daily in the afternoon from 2 to 9 p.m. A model layout is depicted in Fig. 2.10. Simulation of the discharge process indicates that the total number of discharges for one typical week is about 74.

Scenario 2. Proposed discharge process: one dedicated nurse who performs discharges for all units as they ordered. The nurse's shift duration is 8 h, from 1 to 9:30 p.m. with 30 min lunch break, as indicated on the panel layout in Fig. 2.11.

Simulation of this discharge process indicates that the total number of discharges for a similar typical week is only about 60.

This is a rather unexpected result: a newly hired dedicated discharge nurse would make significantly less discharges than nurses in the current process, creating additional discharge backlog and waiting time.

Why does this seemingly good management idea actually create a worse problem?

In the current process, four nurses perform discharges independently. Each nurse has her own path to discharge. If there is a delay with a particular patient in a particular unit, none of the other nurses in other units are impacted.

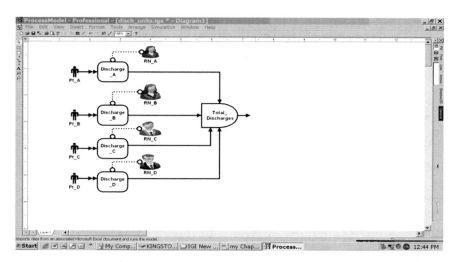

Fig. 2.10 Simulation model layout of current discharge arrangement. Each nurse RN_A, RN_B, RN_C, and RN_D discharges patients individually from each separate unit

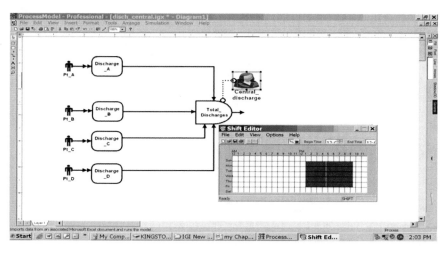

Fig. 2.11 Simulation model layout of the proposed discharge arrangement. One centralized nurse (central discharge) discharges patients from all units. Central discharge nurse shift is from 1 to 8:30 p.m. with 30 min lunch time (light band), as presented on the shift editor panel

With the new centralized discharge process, all discharges from all units form a so-called series of dependent events (Goldratt and Cox 2004, Motwani et al. 1996). In a series of dependent events, a random delay with the discharge of a particular patient would inevitably impact the rest of the patients. In other words, hiring a dedicated resource would likely create a system bottleneck. By definition, a system bottleneck is a resource whose capacity is less than or equal to demand placed on it (Goldratt and Cox 2004).

Based only on averages, a dedicated nurse would be able to perform 80 discharges for a week (4 discharges×4 units×5 days). If each discharge takes on average 30 min, then 40 work-hours for a week is needed to complete them. This makes the capacity of the dedicated discharge nurse equal to demand place on her, i.e., makes her a bottleneck according to the above definition. In reality, the situation is even worse because of the demand variability, and nurse availability that is less than 100% because of additional breaks, meetings, etc.

On the "good" days, when the demand for discharges is less than nurse capacity to perform them (fewer than 4 discharges per unit that take less than 30 min per discharge), the extra capacity cannot be "stored" to serve the next day demand. Such an extra capacity on a particular "good" day is lost. On the other hand, on the "bad days," when demand for discharges exceeds nurse capacity, the unserved demand is not lost; it has to be fulfilled in the next day, forming a backlog (unless there is overtime). Thus, effect of "bad" days is accumulated while effect of "good" days is not.

This illustrates the following fundamental management engineering principles: (1) in a series of dependent events, only the bottleneck defines the throughput of the entire system regardless on the throughput and capacity of nonbottlenecks; (2) an unfulfilled service request backlog (appointments, discharges, document processing, etc.) can exist and remain stable even if the average high variability demand is less than service capacity (Savin 2006).

There are possible solutions that can elevate the bottleneck constraint of the centralized discharge nurse, e.g., scheduling easy discharges with lower variability first (see Sect. 2.4.1.2) or the use of the shared staff. However, a discussion of the techniques for elevating the bottleneck's constraint is beyond the scope of this Brief Series book.

2.4.3 Staffing of Hospital Receiving Center

The hospital receiving center is projected to receive annually 127,139 packages. The time to process one package (screen, scan, store, etc.) could range from 7 to 15 min with the average time of 10 min. The department works Monday to Friday from 8 a.m. to 4:30 p.m., with a 30 min lunch time and two 15 min breaks for each staff member during a typical day. (Total annual number of work days is 255).

It is required to develop a staffing plan for the receiving department with no overtime.

2.4.3.1 Traditional Management Approach (Langabeer 2007)

The daily average number of packages is 127,139/255 = 498.6.

Because each staff member has 30 min off for lunch and two 15 min breaks, the total daily time off is 60 min. Hence each staff member is available for 7.5 h, i.e., the daily availability is 88.2%.

Using the average time of 10 min to handle a package and the available daily work time 7.5 h×60=450 min, each staff member can handle 450/10=45 packages per day. Hence 498.6/45=11.1 staff members are needed to handle the projected work volume without delay and overtime.

2.4.3.2 Management Engineering Approach

A typical traditional approach to calculate the average daily productivity is based on dividing the total available time by the average time per package. Such an approach always results in underestimating of the required resources. This is another illustration of the flaw of averages (Savage 2009; Costa et al. 2003). Notice that in the above calculation the variability range from 7 to 15 min per package is not used at all.

However, it is a rigorous mathematical fact that the average value of a nonlinear function is not equal to the function of the average values of its arguments.

If the low limit of the time range 7 min is used, then only 7.7 staffing FTEs would be needed, while the use of the high limit of 15 min per package results in 16.6 staffing FTEs. The average of these values is not the same as the average staffing 11.1 FTEs based on the average time of 10 min per package.

It is not possible to make a correct staffing calculation without taking into account the frequency of each possible time to handle a package. Only simulation modeling methodology allows one to directly take into account the handling time variability.

The simulation model layout to get the correct staffing in this case is very simple. It is similar to the one discussed in Sect. 2.3.1.3. A time range from 7 to 15 min with the most likely 10 min is the model input in the form of a triangle statistical distribution (in the absence of a more accurate data collection).

Another model input is the daily variable number of packages. While the daily average number of packages is 498.6, the actual number should be of integer type and it varies around this average. It is typically assumed that the daily number of packages is independent of each other, and there are no periods during the year with systematically very high or very low daily package arrivals. Therefore, the daily package arrivals can be represented as a random number from a discrete Poisson distribution with the constant average value parameter 498.6. A number of random samples from this distribution are generated. Each sample consists of 255 random integer numbers (each number represents one work day load of package arrivals). Because the samples are random, only the sample that sums up to 127,139 is picked up as the model input-package arrival schedule.

Simulation results for 51 weeks (8,520 h), Monday to Friday, indicate that 11 FTEs (calculated using the traditional approach based on the average time without variability around the average) would not be able to process the required annual work load of 127,139 packages: the maximum number of packages that could be processed is only 118,744; the 99% confidence interval (CI) of the number of processed packages is from 118,320 to 118,534. Thus, department staffing of 11 FTEs was significantly underestimated.

According to simulation results, staffing of 12.5 FTEs (12 full-time and one 50% part-time) would be needed to process all required annual 127,139 packages (99% CI from 127,137 to 127,139). The daily average staff utilization would be rather healthy, about 88%.

Thus, the traditional management approach would result in chronic department understaffing, and, consequently, underestimated the staffing budget. This, in turn, would result in staff burnout, overstress, and possible increasing in staff turnover.

If budgeting and hiring of the staff is strictly limited, for example, to 11 FTEs, then the department management should provide some additional training or other means that would result in a reduction of the processing time per package and, more importantly, its variability by process standardization.

Simulation modeling easily demonstrates how much reduction of the process time and its variability per package is needed in order to process the required workload with only 11 FTEs.

For example, if the average processing time is reduced to 9 min and the variability is in the range from 7 to 12 min, then 11 FTEs would be able to process practically all package volume (99% CI from 127,136 to 127,139.

Of course, many other operational scenarios are possible to analyze to develop and budget a realistic staffing plan using a simulation model, such as different work shift length for different staff members, part-time shifts, unplanned staff absence, seasonal or quarter variability of package volume, different processing times for different package types, different package handling priorities, and so on.

2.4.4 Staffing of the Unit with Cross-trained Staff

A hospital case management department performs three types of transactions: reservation, urgent admissions, and preregistration. It is expected that the total annual transaction volume is 53,855. About 14% of all transactions are reservation, 32% is urgent admissions, and 52% is preregistration. Case management specialists for urgent admissions and preregistration are cross-trained and could substitute each other if needed. Reservation specialists work independently and they are not involved in performing other type of transactions.

The department works Monday to Friday from 8 a.m. to 4:30 p.m., with a 30 min lunch time and two 15 min breaks for each staff member during a typical day. There are total 255 annual work days. It was estimated that total nonproductive time (unscheduled breaks, distractions, local building trips, personal time, meetings, etc.) for reservation specialists was about 13%, for urgent admission specialists was 15%, and for preregistration specialists was about 22%.

Transaction time data were collected over some representative period and summarized in Table 2.11.

It is needed to develop minimal staffing (FTE) requirement for each transaction type that allows performing the annual transaction volume without overtime (one FTE—full-time equivalent—is assumed to be 8 h).

Table 2.11 Summary of mean and median time per transaction

Transaction type	Time (min)	
	Mean	Median
Reservation	4.1	3.0
Urgent admission	8.3	6.0
Preregistration	4.5	3.0

2.4.4.1 Traditional Management Approach

Similarly to the previous section, the following formulas were applied using the average transaction time and the average staff availability:

Reservation FTE	$53{,}855 \times 14\% \times 4.1$ min$/[255$ days $\times 8$ h $\times 60$ min $\times (1-0.13)] = 0.3$
Urgent admission FTE	$53{,}855 \times 32\% \times 8.3$ min$/[255$ days $\times 8$ h $\times 60$ min $\times (1-0.15)] = 1.4$
Preregistration FTE	$53{,}855 \times 54\% \times 4.5$ min$/[255$ days $\times 8$ h $\times 60$ min $\times (1-0.22)] = 1.4$

Total FTE requirement is 3.1. Notice that cross-trained (shared) staff cannot directly be taken into account by the simple formulas based on average transaction time.

2.4.4.2 Management Engineering Approach

Following the discussion in the previous section, a simulation modeling methodology should be used to accurately assess the staffing requirements. Once again, two main factors contribute to the limitation of the traditional approach: the use of average transaction time rather than transaction time statistical distribution and the inability of simple formulas to take into account the workload of the cross-trained (shared) staff.

Transaction time distributions used in the simulation model are shown in Fig. 2.12 for reservation, urgent admission, and preregistration staff, respectively. It is seen that that these distributions are highly skewed to the larger time side (long distribution tails). Although the mean time per transaction for reservation, urgent admission, and preregistration is relatively small (see Table 2.11), the actual transaction time can be much longer than the mean: up to 18, 38, or 26 min for reservation, urgent admission, and preregistration, respectively. No formulas used in the traditional approach can account for such long time tails. This transaction time variability is a reflection of complexity and/or special case situations for some transactions that require much more staff attention and time. Although such cases are relatively rare, they contribute a lot to the overall staff overload and are difficult to catch-up later.

A simulation model layout is depicted in Fig. 2.13. The layout indicates the overall flow of transaction that is split according to the percentage volume on three particular flows. Dotted lines for urgent admissions and preregistration staff mean that the

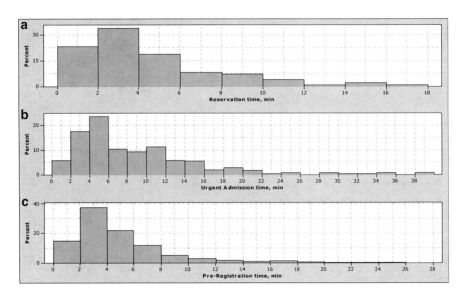

Fig. 2.12 Transaction time distribution for reservation staff (**a**), urgent admission staff (**b**), and preregistration staff (**c**)

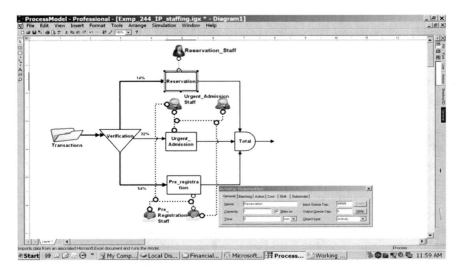

Fig. 2.13 Simulation model layout for cross-trained (shared) staffing of the Case Management Department. *Dotted lines* indicate shared (alternate) work-load assignments for urgent admission and preregistration staff

work-load can be shared between them, i.e., if, for example, the urgent admission staff is completely booked but the preregistration staff is available at the moment, then this staff can pick up the urgent admission cases, and the way around, respectively. Thus, the overall productivity increases because two teams support each other.

Table 2.12 Simulation output for the case management department and required FTE

Transaction type	Simulated annual transaction volume, 99% CI	Required FTE
Reservation	7,460–7,575	0.5
Urgent admission	17,194–17,308	1.5
Preregistration	29,020–29,153	2.5
Total	53,854–53,855	4.5
Target total annual transaction volume	53,855	

Another model input is the daily variable number of transactions. The annual transaction volume is converted into the daily volume using an approach based on Poisson arrival process described in the previous section.

Simulation methodology was based on the changing of the number of FTE for each transaction types until the target total annual transaction volume is hit. This methodology is somewhat similar to design of experiments (DOEs) described in Kolker (2008). Simulation output is presented in Table 2.12.

It follows from this table that practically all annual transaction volume, 53,855, will be performed if total 4.5 FTE are scheduled: 0.5 FTE for reservation, 1.5 FTE for urgent admission, and 2.5 FTE for preregistration. This is significantly higher than traditional estimates of 3.1 FTE. The reason for this discrepancy is already mentioned long transaction times (tails from distribution time) that are not reflected in the mean transaction time. Once again, this is one more illustration of the flaw of averages discussed in details in the previous sects. 2.3.5.3, 2.3.10.2, and 2.4.3.2.

Also, as it was discussed in Sect. 2.4.3.2, it is easy to use the simulation model to find out how much reduction in transaction time variability would be needed if the total staffing is strictly budgeted to, say, 3 FTE or 4 FTE. Simulation modeling (DES models) is indeed indispensable to answering such type of questions (or other questions mentioned at the end of the previous section).

2.4.5 Outpatient Clinic Costs and Staffing: Is Right Staff Used at the Right Time?

An outpatient flu clinic is open during a flu season to provide the flu vaccine shots on a walk-in basis. The clinic stays open from 8 a.m. to 6 p.m.

Giving a shot (including filling out the paper work) takes on average about 8 min but could be in the range from 6 to 10 min. Medical providers (staff) have unpaid 30 min lunch time and two additional paid 15 min short breaks (not overlapped to each other). Total scheduled daily time off for each medical provider is 1 h.

The clinic charges the patient $30 for the flu shot; the clinic's cost of one vaccine dose and supplies is $2. Staffing pay rate is $14.5 per h.

The average weekly patient volumes are collected in Table 2.13.

Table 2.13 The average weekly patient volumes for the outpatient clinic

	Monday	Tuesday	Wednesday	Thursday	Friday	Total	Average weekly arrival rate
8–10 a.m.	17	11	20	20	12	80	8 patients per h
10 a.m.–2 p.m.	44	36	46	62	52	240	12 patients per h
2–4 p.m.	18	17	20	16	19	90	9 patients per h
4–6 p.m.	36	32	37	45	40	190	19 patients per h
Total	115	96	123	143	123	600	

Thus, patient arrival rate is highly variable not only during a typical day but also during days of week.

Clinic's management should decide: how many medical providers are needed to staff the clinic on a typical day, and what is the projected net revenue (on a weekly basis)?

2.4.5.1   Traditional Management Approach

The projected total weekly patient volume (Monday to Friday) is 600. One provider is going to serve on average 60 min/8 min = 7.5 patients per h. Hence $(600/5)/7.5 = 16$ h of staffing time is needed to serve all patients on a typical day. Therefore, two medical providers should be scheduled to staff the clinic on a daily basis. One can be scheduled to work from 8 a.m. to 5 p.m. Another provider can be scheduled to work from 9 a.m. to 6 p.m. Both providers have 8 h shift (1 FTE) and scheduled total 1 h off for lunch and breaks. Practically no (or very short) patient line is expected.

The weekly revenue is going to be $600 \times \$30 = \$18,000$. The weekly labor cost for two providers is $(\$14.5/h \times 8.5\ h \times 5\ days) \times 2 = \$1,232.5$. The total weekly vaccine and supplies costs is $600 \times \$2 = \$1,200$. Hence, the average clinic's weekly net revenue is expected to be $\$18,000 - \$1,232.5 - \$1,200 = \$15,567.5$.

2.4.5.2   Management Engineering Approach

Because of inevitable variability in the daily number of patients coming for the shots, and the variability of the time it takes to give a shot, the actual staffing needs and the actual estimated net revenue will differ significantly from the average values. On top of that, it is observed that some patients leave without a shot if their waiting time is longer than 20 min.

In order to develop a realistic evaluation of clinic performance, the process variability and patients leaving without a shot should be taken into account. This is possible only using simulation modeling of the clinic operations.

The model design and layout is a combination of the models described in Sects. 2.3.3.1 and 2.3.4.3 with patients leaving after waiting more than 20 min (of course, any other numbers and input information can be used).

Simulation input data include one provider working from 8 a.m. to 5 p.m. (1 FTE—full-time equivalent) and another one starting later from 9 a.m. until the end of the day 6 p.m. (1 FTE). On top of scheduled 30 min time off and two 15 min breaks, providers are assumed to spend additionally about 5% of their time (20–25 min) for personal needs, i.e., their practical availability is about 95%.

Simulation reveals that 99% confidence interval (CI) for the weekly number of served patients ranges from only 474 to 476. This is much lower than the expected average value 600. On top of that, 121–123 patients (about 20%) will leave weekly without a shot because of waiting longer than 20 min.

The 99% CI for the weekly net revenue ranges from $12,054 to $12,110. This is much less than the expected average value $15,567.5. Such a significant decreasing of the net revenue results from the inevitable process variability as well as many patients leaving without the shot. Thus, staffing of only two providers is not enough to meet the clinic's performance targets.

The next management step might be to find out how much it would help to increase the net revenue and reduce the number of leaving patients if both providers work, for example, extended hours, from 8 a.m. to 6 p.m. From a traditional management standpoint this should not be needed because 16 h of working time on average should be enough to meet the average patient demand for service time.

Nonetheless, simulation of this scenario indicated that the 99% CI for the number of served patients increased to 513–515, and 83–85 patients left without a shot. The 99% CI for net revenue has become markedly higher than for regular hours, $12,912–$12,969. Certainly, an additional work hour for each provider (overtime) is paid at a 50% higher rate. This results in additional operating expenses (increased labor costs). However, these costs are amply offset by the clinic's higher net revenue because more paying patients are served.

Despite the improved clinic's performance in terms of net revenue, it is still sometimes desired to avoid overtime. Therefore, another option could be, for example, hiring a part-time provider. Suppose, for example, that an additional 0.6 FTE with total 30 min paid time off is placed in the morning shift from 8 a.m. to 1 p.m. How much will it help?

Simulation of this scenario indicated that the number of served patients will be lower than in the previous scenario with the regular staff overtime: 99% CI is only 508–509, and 99% CI net revenue is $12,620–$12,669.

Does this mean that more than additional 0.6 FTE is needed? Not necessarily. What if the same 0.6 FTE is placed in the second shift, from 1 to 6 p.m.? This scenario results in much better performance: 99% CI for the number of served patients is now 557–559, only 38–40 patients left without a shot and 99% CI for net revenue increases to $14,000–$14,061.

Thus, simulation modeling demonstrates that an additional third provider (0.6 FTE) placed in the right shift does help to serve more patients and increase the net revenue despite the higher costs of keeping an additional provider.

Of course, many other scenarios of staffing shifts and the clinic's operation modes are possible to analyze using a simulation model. The clinic's manager might be interested, for example, in knowing what staffing is needed to serve all 600

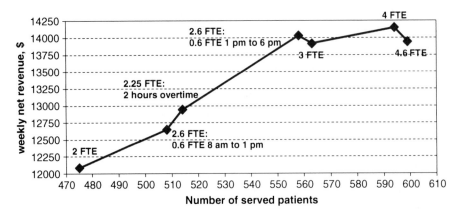

Fig. 2.14 Net revenue vs. the number of served patients and corresponding FTE

projected weekly patient volume. Simulation modeling indicates that practically all weekly patient volume will be served using 4.6 FTE (four full-time FTE and a part-time 0.6 FTE in the second shift). However, the net revenue in this case will drop down to $13,933–$13,948.

A graphical summary that illustrates these and other simulated scenarios is presented in Fig. 2.14.

It follows from Fig. 2.14 that large weekly net revenue (about $14,030) is generated from serving about 558 patients using 2.6 FTEs with 0.6 FTE in the afternoon. However, a little higher net revenue ($14,149) can be obtained using 4 FTE that serve many more patients, about 594. The cost of additional 1.4 FTE is practically offset by the revenue from serving about 36 more patients. Therefore, the latter staffing arrangement with 4 FTE would be preferred. On the other hand, if there is a shortage of providers or the staffing budget is strictly limited, then 2.6 FTE would be enough to generate about the same net revenue.

Notice, that although 4.6 FTE serve practically all patients, about 599–600, the net revenue is lower than that for 4 FTE because in this case the cost of additional 0.6 FTE is not offset by the additional revenue from serving just a few more patients (only about 5–6 more). A similar picture is observed comparing 3 FTE and 2.6 FTE: the cost of additional 0.4 FTE is not offset by the revenue from serving only about five more patients using 3 FTE.

This analysis illustrates an important general trade-off between the number of served patients, cost of resources to serve them, and the net revenue. (The net revenue is defined here as the difference between the revenue generated by served patients and the operational staffing and supplies costs required to providing this service.) The net revenue increases if the revenue growth from serving more patients offsets the growth of staffing and supplies costs (resources) to serve them. However, at some point the growth of the resource costs exceeds the growth of the revenue generated by serving only a few more patients; hence, the net revenue goes down.

Chapter 3
Linear and Probabilistic Resource Optimization and Allocation Problems

Abstract This chapter includes detailed analysis of five problems using the linear and probabilistic resource optimization and allocation methodology: linear optimization of patient service volumes for different service lines; optimal staffing for 24/7 three-shift operations; physician resident scheduling to meet Institute of Medicine (IOM) new restricted resident work hours for day and night shifts; optimized specimen mass screening testing aimed at reducing the overall number of tests per specimen; and the projection of the expected number of patients discharged from the Emergency Department given the time that a patient has already stayed in ED (the use of the concept of the conditional probability of discharge).

Keywords Linear optimization • Objective function • Constraints • Excel solver • Total probability • Conditional probability • Emergency Department • Patient discharge • Physician resident scheduling • Optimal staffing • Specimen mass screening

3.1 Optimization of Patient Service Volumes: Keep or Drop a Service Line?

A hospital is in the process of evaluating the financial viability of three patient service lines. These service lines require the use of five types of hospital resources. The average values of these resources (both per patient and the annual total limits) are presented in Table 3.1. The average net revenue per patient for each service line is also included in Table 3.1.

Zero values mean that this type of resources is not required at all for the corresponding service line.

The hospital management would like to maximize the total net revenue from these services by deciding which services should be offered and in what volumes (a similar problem was suggested by Ozcan 2009).

Table 3.1 Resources required for service lines and total annual limits

Resource type	Service line 1	Service line 2	Service line 3	Total annual limit
Length of stay	3 days/patient	6 days/patient	4 days/patient	19,710 patient-days
Nursing time	3 h/patient	5 h/patient	4.5 h/patient	16,200 h
Interventional radiology	0.5 h/patient	1 h/patient	0	3,000 h
Laboratory procedures	1 h/patient	2 h/patient	3.5 h/patient	6,000 h
Operating rooms	2 h/patient	0	4 h/patient	1,040 h
Net revenue, $	$560/patient	$790/patient	$1,100/patient	Maximize

3.1.1 Traditional Approach

There is not a really good way to address this problem without some quantitative optimization analysis. It appears that it makes sense to increase the patient volume for service line 3 because it has the highest net revenue per patient, followed by volumes for service lines 2 and 1. However, it is practically very difficult to make sure that total annual resource limits will be met without some kind of quantitative analysis.

3.1.2 Management Engineering Approach

The best approach to solving this resource allocation type of problems is linear optimization (LO). Over the past decades LO methodology was widely applied as a powerful resource allocation planning tool. It is now routinely and widely used in engineering, business, finance, and healthcare academic research but less for practical healthcare management.

The general structure of a typical LO model includes (1) the decision variables, i.e., the variables that define the output of the model; (2) the objective function that describes the quantitative goal that should achieved. The objective function includes decision variables and their linear combinations with some fixed parameters; (3) constraints, i.e., the set of linear inequalities and/or equations that restrict the available alternative and/or amount of resources. Constraints also include decision variables and/or their linear combinations; and (4) parameters with defined numerical values that provide the contribution (weight) of each decision variable into the objective function and constraints.

The goal of the LO model is to maximize or minimize the value of an objective function subject to a set of constraints by varying the decision variables. The set of decision variables that meets the goal is the solution of the LO model.

Let X_1, X_2, and X_3 are patient volumes for service lines 1, 2, and 3, accordingly. These are the decision variables for this particular problem. Using parameters from Table 3.1, the objective function (OF) is the total annual net revenue that should be maximized subject to the set of five functional constraints (3.2)–(3.6):

$$OF, \$ = \$560 \times X_1 + \$790 \times X_2 + \$1,100 \times X_3 -> \text{max} \qquad (3.1)$$

Because the total annual amount of resources is limited, the optimal patient volumes are subject to a set of the following constraints:

$$\text{Length of stay: } 3 \times X_1 + 6 \times X_2 + 4 \times X_3 <= 19,710, \qquad (3.2)$$

$$\text{Nursing time: } 3 \times X_1 + 5 \times X_2 + 4.5 \times X_3 <= 16,200, \qquad (3.3)$$

$$\text{Interventional radiology: } 0.5 \times X_1 + 1 \times X_2 + 0 \times X_3 <= 3,000, \qquad (3.4)$$

$$\text{Laboratory procedures:} 1 \times X_1 + 2 \times X_2 + 3.5 \times X_3 <= 6,000, \qquad (3.5)$$

$$\text{Operating rooms: } 2 \times X_1 + 0 \times X_2 + 4 \times X_3 <= 1,040. \qquad (3.6)$$

There are two more types of constraints that should also be included into this particular problem. The decision variables X_1, X_2, and X_3 should obviously be positive (including zero values). In order to exclude possible negative optimal solutions that could be mathematically correct but practically meaningless, a nonnegativity constraint is used, such as

$$X_1, X_2, \text{and } X_3 >= 0. \qquad (3.7)$$

The optimal decision variables should also be of integer type because they represent the number of patients (patient volumes). Sometimes it is possible to keep the patient volume decision variables as real fractional values, and round up the solution to the nearest integer value. But in many situations it is better to add another constraint

$$X_1, X_2, \text{and } X_3 (\text{all integer}). \qquad (3.8)$$

Solution of this problem (and many other more complex LO problems) can easily be obtained using, for example, Add-in tool Solver in Microsoft Excel spreadsheet. Using Solver setup with the checked boxes "Assume Linear Model" and "Assume Nonnegativity", the optimal solution is found to be $X_1 = 520$, $X_2 = 2,740$, and $X_3 = 0$. The maximal annual net revenue is $2,455,800.

This optimal solution means that patient volume for service line 1 should be 520, patient volume for service line 2 should be 2,740, and service line 3 should be dropped entirely! This is a rather unexpected result because service line 3 is the most profitable per patient. Yet it should be dropped; otherwise, some problem constraints will be violated!

The optimal solution means that any other patient volume will result in violation of some constraints. For example, if we still want to keep service line 3 at some level, say 100 (for example, to keep some doctors in practice), then constraints (3.5) Laboratory Procedures and (3.6) Operating rooms will be significantly violated exceeding their corresponding annual limits by 350 h and 400 h; i.e., 100 more patients cannot be served using available resources. The most violation-prone or so-called binding constraints in this problem are constraints (3.4), (3.5), and (3.6). For the optimal solution their values are already equal to the limits of 3,000, 6,000, and 1,040 h, respectively. Therefore adding even a few more patients to service line 3 will result in violation of these constraints.

Certainly, the hospital management could consider adding some additional resources (if they are available) to elevate these constraints, or to reduce some service time per patient (modify parameters). For example, if the average surgical time per patient for service line 3 is reduced from 4 to 1 h and Lab Procedures reduced to 1.5 h, then the optimal solution that satisfies all constraints will be $X_1 = 261$, $X_2 = 2,617$, and $X_3 = 518$. Thus, in this case service line 3 should be kept at the level 518 patients, while service line 1 patient volume should be reduced down to 261 from 520. The maximal annual net revenue is going to be higher, $2,783,390, because the financial contribution of service line 3 into the overall net revenue is higher.

Of course, many other scenarios are possible to evaluate using LO models similar to the methodology of discrete event simulation (DES) modeling presented in Sect. 2.3 and 2.4.

There is, however, an important caveat to keep in mind. It was mentioned that the present formulation of a LO model assumes that all model parameters have defined fixed values, i.e. the optimization problem is deterministic. Despite the fact that the model parameters in Table 3.1 are actually random variables, only their expected (average) values were used in search of the optimal solution. These average values were actually treated as if they are known with certainty (deterministic). This can somewhat limit the validity of LO model results. In deterministic linear optimization models all the information necessary to search the optimal solution should be available before the search begins. On the other hand, uncertainty in parameter values makes not all the information available before the search begins, and some parameters should be modeled as random variables (Sen and Higle 1999). If one or more of the data elements in a LO model are represented by a random variable, a stochastic LO problem arises (SLO). In general, replacing random variable parameters by their average values in LO problem results in solutions that can be structurally different from those provided by corresponding stochastic optimization models (Sen and Higle 1999). The presence of uncertainty affects both feasibility (constraints) and optimality (objective function). This is one more face of the "flaw of averages" discussed in Sects. 2.3.5.3, 2.3.10.2, 2.4.3.2, and 2.4.4.2.

There are two main approaches used to analyze stochastic LO problems (Prekopa 1995; Kall and Mayer 2005). One is based on modeling of recourse (future response). Another one restricts the probability of constraints violation to some prespecified level (problems with probabilistic constraints). However, discussion of these approaches is well beyond the scope of this Springer Brief book.

3.2 Optimization of Clinical Unit Staffing for 24/7 Three-Shift Operations: Is Staffing Cost Minimized?

Scheduling staff is a typical but difficult and time-consuming task in operations management. A typical full-time unit nurse (such as an ICU nurse) usually works 5 days a week with two consecutive rotating days off and rotating shifts. Usually three 8-h shifts per day should be covered. A typical clinical unit has some minimal staffing requirement based on the average shift patient census and assumed nurse to patient ratio. This ratio is based on assessed patient acuity level or external regulations.

It is assumed here that the pay rate is $50/h (base wages and overhead) with 50% pay rate increase for Saturday and Sunday shifts.

For a 7-day week and three shifts per day there are total 21 different schedules possible. These schedules are presented in Table 3.2 along with the average shift census and the minimal staff demands for each shift assuming nurse to patient ratio 1:2 for all shifts (only Monday to Thursday are shown in this table. Friday, Saturday, and Sunday are structured similarly but not shown here to save space).

The management is supposed to develop a staffing schedule to meet the minimum coverage for each day and shift with five work days and two consecutive days off for each staff member in such a way that the total weekly staffing cost is minimized.

3.2.1 Traditional Approach

Typically the staffing demand per shift is estimated using the historical average shift census and assumed nurse to patient ratio based on assessed acuity or external regulations. This approach cannot, of course, guarantee that all staffing shift constraints are met because the focus is usually on reducing the total cost of staffing. This practice inevitably results in daily tweaking of staff, requests to work overtime, and ultimately eating up all savings.

3.2.2 Management Engineering Approach

According to the general approach to solving linear optimization (LO) problems, it is required to define (1) decision variables, (2) objective function, (3) constraints, and (4) model parameters.

Decision variables for this problem is the number of nurses, X_s ($s = 1,\ldots, 21$) assigned to each of $s = 21$ schedules.

Objective function is the total weekly staffing cost for all shifts, C, that should be minimized by placing right staff in the right shift.

Table 3.2 Binary index variable I_{sds} for $s = 21$ schedules and days and shifts: 1—on shift, 0—off shift

Schedule s	Days off	Mon_shift_1	Mon_shift_2	Mon_shift_3	Tue_shift_1	Tue_shift_2	Tue_shift_3	Wed_shift_1	Wed_shift_2	Wed_shift_3	Thu_shift_1	Thu_shift_2	Thu_shift_3
1	Sat, Sun	1	0	0	1	0	0	1	0	0	1	0	0
2	Sat, Sun	0	1	0	0	1	0	0	1	0	0	1	0
3	Sat, Sun	0	0	1	0	0	1	0	0	1	0	0	1
4	Sun, Mon	0	0	0	1	0	0	1	0	0	1	0	0
5	Sun, Mon	0	0	0	0	1	0	0	1	0	0	1	0
6	Sun, Mon	0	0	0	0	0	1	0	0	1	0	0	1
7	Mon, Tue	0	0	0	0	0	0	1	0	0	1	0	0
8	Mon, Tue	0	0	0	0	0	0	0	1	0	0	1	0
9	Mon, Tue	0	0	0	0	0	0	0	0	1	0	0	1
10	Tue, Wed	1	0	0	0	0	0	0	0	0	1	0	0
11	Tue, Wed	0	1	0	0	0	0	0	0	0	0	1	0
12	Tue, Wed	0	0	1	0	0	0	0	0	0	0	0	1
13	Wed, Thu	1	0	0	1	0	0	0	0	0	0	0	0
14	Wed, Thu	0	1	0	0	1	0	0	0	0	0	0	0
15	Wed, Thu	0	0	1	0	0	1	0	0	0	0	0	0
16	Thu, Fri	1	0	0	1	0	0	1	0	0	0	0	0
17	Thu, Fri	0	1	0	0	1	0	0	1	0	0	0	0
18	Thu, Fri	0	0	1	0	0	1	0	0	1	0	0	0
19	Fri, Sat	1	0	0	1	0	0	1	0	0	1	0	0
20	Fri, Sat	0	1	0	0	1	0	0	1	0	0	1	0
21	Fri, Sat	0	0	1	0	0	1	0	0	1	0	0	1
	Average census per shift	10	24	20	21	18	15	8	15	24	18	20	21
	Minimal staff demand per day and shift, $N_{ds, MinDemand}$	5	12	10	10	9	7	4	7	12	9	10	10

Nurse to patient ratio is assumed to be 1:2 for all shifts

Let L_{ds} be the shift length for the particular day and shift (ds) (assumed here to be 8 h for all days and shifts), and P_{ds} is the pay rate per hour, which is \$50/h including overhead for Monday to Friday day shifts (ds), ds = 1,..., 15 and 1.5 times that for Saturday and Sunday shifts, i.e., \$75/h for ds = 16,..., 21. (Mon_shift_1 is indexed as ds = 1, Mon_shift_2 is ds = 2, and so on to Sun_shift_3, which is indexed as ds = 21).

Let N_{ds} be the number of nurses with different schedules (s = 1,..., 21) assigned to the particular day and shift (ds = 1,..., 21). Then

$$N_{ds} = \sum_{s=1}^{21} X_s I_{s,ds} \quad \text{for each} \quad ds = 1,...,21.$$

The binary index variable $I_{s,ds}$ is equal to 0 or 1 as indicated in Table 3.2 (only Monday to Thursday shifts shown to save space). If $I_{s,ds}$ is 0 then contribution of the decision variable X_s to the total sum is also zero; i.e., nurses with schedule s are not assigned for the particular day and shift, ds. If $I_{s,ds}$ is equal to 1 then the decision variable X_s is counted in the total sum; i.e., nurses with schedule s are assigned for the particular day and shift ds.

The total weekly nursing cost objective function to be minimized is calculated as

$$C = \sum_{ds=1}^{21} N_{ds} P_{ds} L_{ds} = \sum_{ds=1}^{21} P_{ds} L_{ds} \sum_{s=1}^{21} X_s I_{s,ds} \Rightarrow \min.$$

Constraints for decision variables X_s (s = 1,..., 21) are the minimal total staff demand for each day and shift, $N_{ds, \text{Min Demand}}$. These values are calculated using the average census per shift (indicated in Table 3.2) and the nurse to patient ratio (1:2 in this case). $N_{ds, \text{Min Demand}}$ values are indicated in the last row of Table 3.2. Thus, the optimal solution X_s (s = 1,..., 21) must satisfy the inequalities

$$N_{ds} = \sum_{s=1}^{21} X_s I_{s,ds} \geq N_{ds,\text{MinDemand}}, \quad \text{for each} \quad ds = 1,...,21.$$

Other constraints for decision variables X_s (s = 1,..., 21) are their nonnegativity and integer values. Model parameters are the nurse to patient ratio, the average census per shift, as well as pay rates and shift length.

This completes the linear optimization model setup. Solution to this problem can easily be found using widely available Microsoft Excel Solver.

Specific details of entering this model in Microsoft Excel Solver cells are beyond the scope of this book. They can easily be found elsewhere, for example, in Winston (2004).

The optimal solution is presented in Table 3.3 for Monday to Thursday shifts (Friday to Sunday are not shown to save space). Thus, in order to implement the schedule with three daily shifts and rotating consecutive 2 days off the total required nursing pool size should be $N_{tot} = \sum_{s=1}^{21} X_s = 36$. Table 3.3 provides information for

Table 3.3 Optimal solution to scheduling problem (Friday to Sunday are not shown)

Schedule s = 1,…,21	Days off	Optimized Solution, Xs	Mon_shift_1	Mon_shift_2	Mon_shift_3	Tue_shift_1	Tue_shift_2	Tue_shift_3	Wed_shift_1	Wed_shift_2	Wed_shift_3	Thu_shift_1	Thu_shift_2	Thu_shift_3
1	Sat, Sun	3	3			3			3			3		
2	Sat, Sun	7		7			7			7			7	
3	Sat, Sun	8			8			8			8			8
4	Sun, Mon	4				4			4			4		
5	Sun, Mon	1					1			1			1	
6	Sun, Mon	1						1			1			1
7	Mon, Tue	1							1			1		
8	Mon, Tue	0												
9	Mon, Tue	1									1			1
10	Tue, Wed	0												
11	Tue, Wed	2		2									2	
12	Tue, Wed	0												
13	Wed, Thu	2	2			2								
14	Wed, Thu	0												
15	Wed, Thu	0												
16	Thu, Fri	0												
17	Thu, Fri	3		3			3			3				
18	Thu, Fri	2			2			2			2			
19	Fri, Sat	1	1			1			1			1		
20	Fri, Sat	0												
21	Fri, Sat	0												
Actual staff demand per shift, N_{ds}			6	12	10	10	11	11	9	11	12	9	10	10
Minimal staff demand per shift, $N_{ds,\,Min\ Demand}$			5	12	10	10	9	7	4	7	12	9	10	10
Excess of actual demand over minimal demand			1	0	0	0	2	4	5	4	0	0	0	0

the number of nurses that should be assigned on the particular day for the particular shift. For example, for shift 1 on Monday three nurses should be assigned with schedule $s = 1$ (Sat and Sun off), two nurses should be assigned with schedule $s = 13$ (Wed and Thu off), and one nurse should be assigned with schedule $s = 19$ (Fri and Sat off). Thus, total six nurses should be assigned for shift 1 on Monday.

Similarly, say, for shift 3 on Wednesday eight nurses should be assigned with schedule $s = 3$ (Sat and Sun off), one nurse with schedule $s = 6$ (Sun and Mon off), one nurse with schedule $s = 9$ (Mon and Tue off), and two nurses with schedule $s = 18$ (Thu and Fri off). Total staffing for shift 3 on Wednesday is 12. The total weekly nursing cost for the optimal staffing is $77,800.

Many other scenarios are possible to analyze using LO methodology. For example, nurse to patient ratio and pay rate could be different from shift to shift and day to day, or part-time staff can be added to the schedule. It is possible to set up a LO problem aimed at maximizing nurse satisfaction using employees preference score for a chosen shift (McLaughlin and Hays 2008).

At the same time, there is an important caveat for LO model results interpretation similar to that mentioned in the previous section. Census values are treated here as fixed parameters, although these numbers are just the averages over the variable census described by some statistical distributions. Therefore this problem is, strictly speaking, a stochastic linear optimization problem. However, describing and using stochastic linear optimization is beyond the scope of this book. As it was mentioned in the previous section, stochastic linear optimization coverage is available, for example, in Prekopa (1995), Kall and Mayer (2005).

3.3   Resident Physician Restricted Work Hours: Optimal Scheduling to Meet the Institute of Medicine New Workload Recommendations

Long working hours and sleep deprivation have been typical for resident physician training programs in the US hospitals for a long time. However, there is abundant scientific evidence that links fatigue with deficits in human performance, accidents, and errors in the various safety-sensitive industries including medicine (Blum et al. 2011). In 2009, the Institute of Medicine (IOM) published a detailed report that examined the scientific evidence linking resident physician sleep deprivation with clinical performance deficits and medical errors (Ulmer et al. 2009). In this report new limits are recommended, in particular, on resident physician work hours and workload and other aspects of residency training: limit resident physician work hours to 12–16 h maximum shifts, and a minimum of 10 h off duty should be scheduled between shifts. In addition to these limits "...At least one 24-hour off duty period must be provided per seven-day period without averaging; one additional (consecutive) 24-hour period off duty must be provided to ensure at least one continuous 48-hour period off duty per month", and "...Night float or night shift duty must not exceed four consecutive nights and must be followed by a minimum of 48 continuous hours off duty after three or four consecutive nights."

Suppose that a hospital critical care unit (ICU) wants to implement these recommendations for its physician resident program. It is known that for a typical 12-h weekday shift at least four residents should be scheduled, while on weekend (Saturdays and Sundays) at least three residents should be available.

For night shifts, at least three residents should available on weekdays and at least two residents should be available for Saturdays and Sundays.

It is needed to determine the minimal total number of residents and to develop a schedule that meets the new IOM restricted hours recommendation and provides the minimal required daily coverage of residents.

3.3.1 Traditional Management Approach

Typically, schedules are developed manually or using some scheduling software. Making the schedule manually is not practical. Using scheduling software can help but most of them provide coverage but not optimized scheduling in terms of minimal staffing subject to constraints.

3.3.2 Management Engineering Approach

It appears that linear optimization (LO) methodology would be the most appropriate approach in this case. It allows determining the minimal number of residents subject to a set of constraints indicated in the problem description.

According to the general approach to solving LO problems, it is required to define (1) decision variables, (2) objective function, (3) constraints, and (4) model parameters.

3.3.3 Day Time Scheduling

LO approach to scheduling in this case is somewhat similar to the one used in Sect. 3.2. However, this scheduling problem is different because the IOM recommendations require actually the development of the monthly (4 weeks) schedule rather than the weekly one in order to implement at least one monthly 2 day-off period (golden weekend) along with at least one weekly day off for all other weeks in this month. These requirements result in 196 possible schedules! Indeed, it is possible to make seven schedules with any two consecutive days off for any one of

4 weeks for the month along with the other weeks with a single day off, say, on Monday, i.e., $7 \times 4 = 28$ schedules for 1 month. This pattern can be repeated for the weeks with other single days off, i.e., Tuesday, Wednesday, and so on. Thus, there are total $28 \times 7 = 196$ schedules possible.

Decision variable for this problem is the number of residents, X_s, assigned to each of $s = 196$ possible schedules. Objective function is the total number of residents with all schedules N_{tot} that should be minimized, i.e.,

$$N_{tot} = \sum_{s=1}^{196} X_s \Rightarrow \min.$$

Constraints are the minimal total staff demand for each of the seven days of each of 4 weeks (dw = 1,..., 28). Monday week 1 (Mon1) is indexed dw = 1, Tuesday week 1 (Tue1) is indexed dw = 2, and so on to Sunday week 4 (Sun4) indexed as dw = 28. It is assumed in this example that $N_{dw, min} = 4$ for each weekday (Monday to Friday for each of 4 weeks), and $N_{dw, min} = 3$ for each of four weekends (Saturday and Sunday for each week). Thus, the optimal solution X_s ($s = 1,...,$ 196) must satisfy the inequalities

$$N_{dw} = \sum_{s=1}^{196} X_s I_{s,dw} \geq N_{dw,min}, \quad \text{for each} \quad dw = 1,...,28.$$

The values of binary index variable $I_{s, dw} = 0$ or $I_{s, dw} = 1$ are indicated in Table 3.4. Only first $s = 7 \times 6 = 42$ schedules for the first 2 weeks are shown in this table to save space. Mon234 designates Monday off on weeks 2, 3, and 4; Mon134 designates Monday off on weeks 1, 3, and 4; and so on, respectively.

If $I_{s, dw}$ is equal to 0 then contribution of the decision variable X_s to the total sum is also zero; i.e., residents with schedule s are not assigned for the particular day of the week, dw. If $I_{s, dw}$ is equal to 1 then the decision variable X_s is counted in the total sum; i.e., residents with schedule s are assigned for the particular day of the particular week, dw.

Other constraints for decision variables X_s are their nonnegativity and integer values. There are no parameters in this particular problem. This completes the linear optimization model setup.

Solution to this problem can be found using widely available Microsoft Excel Solver or IBM ILOG CPLEX Solver software (www.aimms.com/licensing/free-licenses).

The Excel Solver optimal solution is presented in Table 3.5. Total staffing pool should be five residents. On Monday week 1 one resident should be scheduled with weekly days off on Tuesdays on weeks 1, 2, and 3, and a monthly two consecutive days off on Thursday and Friday on week 4. Another resident should be scheduled with weekly days off on Wednesdays on weeks 2, 3, and 4 and a monthly two consecutive days off on Wednesday and Thursday on week 1, and so on. Blank cells indicate that no resident with the particular schedule should be scheduled on that day and week.

Table 3.4 Binary index variable for day time shift schedule development: 1—on shift, 0—off shift (first 42 schedules for 2 weeks are shown to save space)

Schedule, s	Weekly day off	Monthly Golden Days off	Mon1	Tue1	Wed1	Thu1	Fri1	Sat1	Sun1	Mon2	Tue2	Wed2	Thu2	Fri2	Sat2	Sun2
1	Mon234	Sun1, Mon1	0	1	1	1	1	1	0	0	1	1	1	1	1	1
2	Mon234	Mon1, Tue1	0	0	1	1	1	1	1	0	1	1	1	1	1	1
3	Mon234	Tue1, Wed1	1	0	0	1	1	1	1	0	1	1	1	1	1	1
4	Mon234	Wed1, Thu1	1	1	0	0	1	1	1	0	1	1	1	1	1	1
5	Mon234	Thu1, Fri1	1	1	1	0	0	1	1	0	1	1	1	1	1	1
6	Mon234	Fri1, Sat1	1	1	1	1	0	0	1	0	1	1	1	1	1	1
7	Mon234	Sat1, Sun1	1	1	1	1	1	0	0	0	1	1	1	1	1	1
8	Mon134	Sun2, Mon2	0	1	1	1	1	1	1	0	1	1	1	1	1	0
9	Mon134	Mon2, Tue2	0	1	1	1	1	1	1	0	0	1	1	1	1	1
10	Mon134	Tue2, Wed2	0	1	1	1	1	1	1	1	0	0	1	1	1	1
11	Mon134	Wed2, Thu2	0	1	1	1	1	1	1	1	1	0	0	1	1	1
12	Mon134	Thu2, Fri2	0	1	1	1	1	1	1	1	1	1	0	0	1	1
13	Mon134	Fri2, Sat2	0	1	1	1	1	1	1	1	1	1	1	0	0	1
14	Mon134	Sat2, Sun2	0	1	1	1	1	1	1	1	1	1	1	1	0	0
15	Mon124	Sun3, Mon3	0	1	1	1	1	1	1	0	1	1	1	1	1	1
16	Mon124	Mon3, Tue3	0	1	1	1	1	1	1	0	1	1	1	1	1	1
17	Mon124	Tue3, Wed3	0	1	1	1	1	1	1	0	1	1	1	1	1	1
18	Mon124	Wed3, Thu3	0	1	1	1	1	1	1	0	1	1	1	1	1	1
19	Mon124	Thu3, Fri3	0	1	1	1	1	1	1	0	1	1	1	1	1	1
20	Mon124	Fri3, Sat3	0	1	1	1	1	1	1	0	1	1	1	1	1	1
21	Mon124	Sat3, Sun3	0	1	1	1	1	1	1	0	1	1	1	1	1	1
22	Mon123	Sun4, Mon4	0	1	1	1	1	1	1	0	1	1	1	1	1	1
23	Mon123	Mon4, Tue4	0	1	1	1	1	1	1	0	1	1	1	1	1	1
24	Mon123	Tue4, Wed4	0	1	1	1	1	1	1	0	1	1	1	1	1	1
25	Mon123	Wed4, Thu4	0	1	1	1	1	1	1	0	1	1	1	1	1	1
26	Mon123	Thu4, Fri4	0	1	1	1	1	1	1	0	1	1	1	1	1	1

#														
27	Mon123	Fri4, Sat4	0	0	1	1	1	1	1	1	1	1	1	1
28	Mon123	Sat4, Sun4	0	0	1	1	1	1	1	1	1	1	1	1
29	Tue234	Sun1, Mon1	0	0	1	1	1	0	1	0	1	1	1	1
30	Tue234	Mon1, Tue1	0	0	0	1	1	0	1	0	1	1	1	1
31	Tue234	Tue1, Wed1	1	1	0	0	1	1	1	1	0	1	1	1
32	Tue234	Wed1, Thu1	1	1	0	0	0	1	1	1	0	1	1	1
33	Tue234	Thu1, Fri1	1	1	1	0	0	1	0	0	1	1	1	1
34	Tue234	Fri1, Sat1	1	1	1	1	0	1	0	0	1	1	1	1
35	Tue234	Sat1, Sun1	1	1	1	1	1	1	1	1	1	1	1	1
36	Tue134	Sun2, Mon2	1	1	0	0	0	0	0	1	1	1	1	1
37	Tue134	Mon2, Tue2	1	1	0	0	0	0	1	1	1	1	1	1
38	Tue134	Tue2, Wed2	1	1	0	0	0	0	1	1	0	1	1	1
39	Tue134	Wed2, Thu2	1	1	0	0	0	0	1	1	0	0	1	1
40	Tue134	Thu2, Fri2	1	1	0	0	1	1	1	0	1	1	0	1
41	Tue134	Fri2, Sat2	1	1	0	0	1	1	1	0	1	1	0	0
42	Tue134	Sat2, Sun2	1	1	0	0	1	1	1	0	1	1	1	0

Table 3.5 Optimal solution to day time resident scheduling problem

Schedule, s	54	60	152	154	182	Total	Constraint
Weekly days off	Tue123	Wed234	Sat134	Sat134	Sun134		
Monthly Golden days off	Thu4, Fri4	Wed1, Thu1	Thu2, Fri2	Sat2, Sun2	Sat2, Sun2		
Mon1	1	1	1	1	1	5	4
Tue1		1	1	1	1	4	4
Wed1	1		1	1	1	4	4
Thu1	1		1	1	1	4	4
Fri1	1	1	1	1	1	5	4
Sat1	1	1			1	3	3
Sun1	1	1	1	1		4	3
Mon2	1	1	1	1	1	5	4
Tue2		1	1	1	1	4	4
Wed2	1		1	1	1	4	4
Thu2	1	1		1	1	4	4
Fri2	1	1		1	1	4	4
Sat2	1	1	1			3	3
Sun2	1	1	1			3	3
Mon3	1	1	1	1	1	5	4
Tue3		1	1	1	1	4	4
Wed3	1		1	1	1	4	4
Thu3	1	1	1	1	1	5	4
Fri3	1	1	1	1	1	5	4
Sat3	1	1			1	3	3
Sun3	1	1	1	1		4	3
Mon4	1	1	1	1	1	5	4
Tue4	1	1	1	1	1	5	4
Wed4	1		1	1	1	4	4
Thu4		1	1	1	1	4	4
Fri4		1	1	1	1	4	4
Sat4	1	1			1	3	3
Sun4	1	1	1	1		4	3

3.3.4 Night Time Scheduling

The overall LO approach for the development an optimal night shift schedule is similar to the one described above. However, there is a complication due to the requirement of the cycle of two consecutive days off after three or four consecutive nights. Total cycle length (on–off) is less than 7-day week. Therefore, there are no fixed days off for different weeks (as it was in the previous day time schedule development). Assuming here four consecutive work nights, a pattern with two consecutive days off will be sliding, and the total cycle (on–off) is repeated every 6 full weeks. The binary index variables (on–off) for $s=6$ schedules are presented in Table 3.6.

Table 3.6 Binary index variables for night time schedule development: 1—on shift, 0—off shift

Week	Schedule, s →	1	2	3	4	5	6
↓	Two sliding consecutive days off starting on week 1 →	Fri1, Sat1	Sat1, Sun1	Mon1, Tue1	Tue1, Wed1	Wed1, Thu1	Thu1, Fri1
1	Mon1	1	1	0	1	1	1
1	Tue1	1	1	0	0	1	1
1	Wed1	1	1	1	0	0	1
1	Thu1	1	1	1	1	0	0
1	Fri1	0	1	1	1	1	0
1	Sat1	0	0	1	1	1	1
1	Sun1	1	0	0	1	1	1
2	Mon2	1	1	0	0	1	1
2	Tue2	1	1	1	0	0	1
2	Wed2	1	1	1	1	0	0
2	Thu2	0	1	1	1	1	0
2	Fri2	0	0	1	1	1	1
2	Sat2	1	0	0	1	1	1
2	Sun2	1	1	0	0	1	1
3	Mon3	1	1	1	0	0	1
3	Tue3	1	1	1	1	0	0
3	Wed3	0	1	1	1	1	0
3	Thu3	0	0	1	1	1	1
3	Fri3	1	0	0	1	1	1
3	Sat3	1	1	0	0	1	1
3	Sun3	1	1	1	0	0	1
4	Mon4	1	1	1	1	0	0
4	Tue4	0	1	1	1	1	0
4	Wed4	0	0	1	1	1	1
4	Thu4	1	0	0	1	1	1

(continued)

Table 3.6 (continued)

Week	Schedule, $s \rightarrow$	1	2	3	4	5	6
4	Fri4	1	1	0	0	1	1
4	Sat4	1	1	1	0	0	1
4	Sun4	0	1	1	1	0	0
5	Mon5	0	0	1	1	1	0
5	Tue5	1	0	0	1	1	1
5	Wed5	1	1	0	0	1	1
5	Thu5	1	1	1	0	0	1
5	Fri5	1	1	1	0	0	1
5	Sat5	0	1	1	1	0	0
5	Sun5	0	0	1	1	1	0
6	Mon6	1	0	0	1	1	1
6	Tue6	1	1	0	1	1	1
6	Wed6	1	1	1	0	1	1
6	Thu6	1	1	1	0	0	1
6	Fri6	0	1	1	1	0	0
6	Sat6	0	0	1	1	0	0
6	Sun6	0	0	1	1	1	1

There are only six different schedules in this case that start on week 1. For example, schedule 1 starts with four consecutive work nights on week 1 (Mon1 to Thu1) with two consecutive days off (Fri1 and Sat1). Then there is another string of four work nights (Sun1 to Wed2) with another two days off (Thu2 and Fri2), and so on. Schedule 2 starts on week 1 with Mon1 day off, which is consecutive to Sun6 day off because the entire pattern is repeated every 6 weeks, and so on.

The optimal solution that minimizes the total number of residents assigned to all schedules subject to a set of constraints (minimal of three residents on Mondays to Fridays and minimal of two residents for Saturdays and Sundays) was obtained using Microsoft Excel Solver. It is presented in Table 3.7.

It follows from this table that, for example, three residents should be scheduled on Monday, week 1 (Mon1): one resident starting on Monday with Friday and Saturday days off this week (Fri1 and Sat1), and two residents starting on Monday (Mon1) with Wednesday and Thursday days off this week (Wed1 and Thu1). Other schedules are interpreted similarly.

Thus, LO is indeed a very powerful and resourceful methodology for developing rather complex optimized time coverage schedules.

Notice that DES methodology can also be used for developing optimized schedules, as it was demonstrated in Sect. 2.4. However, DES scheduling is more appropriate for dynamic supply and demand balance problems, i.e., for problems with random processes when scheduled staffing directly depends on patient or transaction volumes (workload).

On the other hand, LO can be preferred if little or no randomness is present, and the primary goal is the minimal staffing needed for continuous time (shift) coverage, such as the number of nurses and residents (or attending physicians) that should always be available for the specific time length or shift due to safety or legal regulations regardless of their actual workload.

3.4   Optimized Pooled Screening Testing

The US Center for Disease Control and Prevention (CDC) has revised its recommendations for screening for human immunodeficiency virus (HIV) and now recommends HIV screening for all patients aged 13–64 years in all healthcare settings, including hospital emergency departments, urgent care clinics, inpatient services, sexually transmitted disease clinics, tuberculosis clinics, and primary care offices (Armstrong and Taege 2007; Bozzette 2005).

A large testing laboratory is staffed and equipped to the testing capacity of 60 HIV specimens per day. Due to new CDC recommendations the specimen daily volume has increased to about 100 per day. This results in a testing backlog and frequent staff overtime.

The management is eager to increase the testing capacity in order to reduce the backlog and staff overtime.

Table 3.7 Optimal solution to night time resident scheduling problem

Week	Schedule, s → Two sliding consecutive days off starting on week 1 →	1 Fri1, Sat1	2 Sat1, Sun1	3 Mon1, Tue1	4 Tue1, Wed1	5 Wed1 Thu1	6 Thu1, Fri1	Total residents	Constraints
1	Mon1	1				2		3	3
1	Tue1	1	1			2		4	3
1	Wed1	1	1	1				3	3
1	Thu1	1	1	1				3	3
1	Fri1		1	1		2		4	3
1	Sat1			1		2		3	2
1	Sun1	1				2		3	2
2	Mon2	1	1			2		4	3
2	Tue2	1	1	1				3	3
2	Wed2	1	1	1				3	3
2	Thu2		1	1		2		4	3
2	Fri2			1		2		3	3
2	Sat2	1				2		3	2
2	Sun2	1	1			2		4	2
3	Mon3	1	1	1				3	3
3	Tue3	1	1	1				3	3
3	Wed3		1	1		2		4	3
3	Thu3			1		2		3	3
3	Fri3	1				2		3	3
3	Sat3	1	1			2		4	2

3	Sun3	1	1	1		3	2
4	Mon4	1	1	1		3	3
4	Tue4		1	1	2	4	3
4	Wed4	1		1	2	3	3
4	Thu4	1	1		2	3	3
4	Fri4	1	1		2	4	3
4	Sat4	1	1	1		3	2
4	Sun4		1	1		3	2
5	Mon5			1	2	4	3
5	Tue5	1		1	2	3	3
5	Wed5	1	1		2	3	3
5	Thu5	1	1		2	4	3
5	Fri5	1	1	1		3	3
5	Sat5		1	1		3	2
5	Sun5			1	2	4	2
6	Mon6	1		1	2	3	3
6	Tue6	1	1		2	3	3
6	Wed6	1	1		2	4	3
6	Thu6	1	1	1		3	3
6	Fri6		1	1		3	3
6	Sat6			1	2	4	2
6	Sun6			1	2	3	2

3.4.1   Traditional Management Approach

Because specimen testing is largely automated and must follow a standard operating procedure, testing time per specimen cannot be much reduced. The only option is budgeting additional laboratory staffing and equipment even though the budget is tight, and might not be approved in full.

3.4.2   Management Engineering Approach

The specimen testing capacity cannot be directly increased due to budgeting issues. However, it is possible to reduce the expected overall number of tests per specimen, thereby increasing the overall laboratory capacity.

Indeed, in the current testing arrangement one specimen requires one test (assuming that there is no rework for the same specimen). Thus, the daily workload of 100 specimens requires 100 individual tests that are beyond current laboratory capacity.

However, if prevalence of the disease to test for is low, then most tests come back negative. Therefore, a combined batch of samples pooled together will frequently result in a negative test. A negative test for the batch allows one to declare each individual specimen used to make up this combined batch negative as well, using a single test. If a batch is positive then each individual specimen used to make up this batch should be retested to identify a positive specimen.

There is a trade-off between the overall reduction of the number of tests if the batch is negative and additional retesting if the batch is positive. The problem is identifying an optimal batch size that results in the overall reduction of the number of required tests compared to the original arrangement for testing each individual specimen.

It is assumed that test results of each sample are independent events. The expected total number of tests is the probability of a positive batch times the number of necessary tests (*batch size* +1) plus the probability of a negative batch times 1.

Let n be a batch size and let P be the probability that each individual specimen is tested positive. The expected number of tests per specimen, N, is going to be (Saraniti 2006)

$$N = \frac{[1-(1-P)^n] \times (n+1) + (1-P)^n}{n} = 1-(1-P)^n + \frac{1}{n}.$$

Reduction of the number of tests compared to the current arrangement of one test per specimen is possible only if N is less than 1. In order for this to happen, the probability of a positive specimen should satisfy the inequality condition

$$P < 1 - n^{-\frac{1}{n}}.$$

The maximum of the right-hand side for this inequality for integer numbers is about 0.306 for $n=3$. Therefore, a reduction of the number of tests per specimen is theoretically possible only if the probability of a positive specimen (disease prevalence), P, is less than about 30%. For each P in this range N has its minimal value. If $12.4\% < P < 30.6\%$ then the optimal batch size which minimizes N is exactly $n = 3$. If $P < 11.1\%$ then the optimal batch size which minimizes N can be approximated by the formula (rounded to the nearest integer number)

$$n \cong \frac{1}{\sqrt{P}} + 0.5.$$

For example, according to a CDC report (2008), the HIV prevalence in the US population at the end of 2006 was about $P = 0.447\%$ with 95% CI from 0.427% to 0.468% (without breaking down by risk groups). Using the above formulas, the optimal batch size for this prevalence is 15 and the expected number of tests per specimen is about $N = 0.13$. This gives 87% reduction of the number of tests per specimen; instead of 100 required daily specimen tests only about $100 \times 0.13 \cong 13$ tests (!) would be needed.

In high risk patient group the probability of a positive test (prevalence) is much higher, e.g., up to $P = 10\%$. In this case the optimal batch size is 4, and the number of tests per specimen is about 0.594. However, even in this case only about $100 \times 0.594 \cong 59$ daily tests are needed. This is within the current laboratory capacity of 60 daily tests.

In practical implementation of this technique some additional factors should be taken into account. A particular concern is batch dilution due to sample pooling that could result in reduced tests' analytic specificity and/or sensitivity. This issue has been addressed in the literature (Saraniti 2006).

On the other hand, the basic principle of pooled specimens testing can be further enhanced under certain circumstances by two variations: sorting and multistage testing. Sorting patient specimens into high and low risk groups allows for additional savings when easily identifiable high risk groups have a much greater prevalence than larger low risk groups.

In multistage testing, positive batches are rearranged into new smaller batches, which are then retested (instead of individual sample retesting). This approach is most efficient if prevalence is very low and the analytic sensitivity loss from large pools is minimal (Saraniti 2006).

Notice that this basic test per specimen reduction methodology can be applied to mass testing of any fluids/specimens.

Thus, management engineering demonstrates how to get done more by doing less through smarter and more efficient managing of the available resources.

3.5   Projected Number of Patients Discharged from ED

It was demonstrated in Sect. 2.3.11 that over-improvement in an upstream unit, such as ED, can negatively impact the performance of downstream units, such as Operating Rooms and ICU. This happens because the capacity of the downstream units is often not enough to handle a patient volume increase coming from ED, especially if that increase was not anticipated with sufficient lead time. In order to improve the preparedness of the downstream inpatient units to handle patient volume coming from ED, it would be helpful to predict the number of patients that are expected to be discharged from ED in the next period of time and admitted as inpatients.

This information would give a sufficient lead time to the downstream units' staff to expedite cleaning and making the needed beds available or to consider discharging or transferring some "old" patients to make the appropriate room for the new incoming patients.

3.5.1 Traditional Management Approach

Typical short-term predictions are virtually always made through physician or nurse assessment supplemented by available patient status report. Physicians and nurses are required to assimilate many different pieces of information in attempt to estimate a discharge time. It is not surprising that manually integrating this information is a very difficult task. This makes prediction of discharges subjective and inaccurate (Littig and Isken 2007; Fuhs et al. 1979). Therefore, inpatient units are often informed about the decision to discharge from ED and admit the patient only after the fact when the decision is already made.

3.5.2 Management Engineering Approach

Typically the historical data on the average length of stay (ALOS) for Emergency Severity Index (ESI) is collected in the ED information system. However, the average LOS alone is not enough to make any predictions for future patient discharges because, as it was discussed in the previous sections, very different LOS distributions could have the same (or close) average values.

The problem of anticipated discharges can be addressed using the concept of conditional probability. Given LOS probability density distribution for each ESI, $f_{ESI}(T)$, and given the patient current length of stay, T, what is the probability of patient discharge, q, in the next time period, t, following the time T?

Thus, if the patient was not discharged at time T (current LOS $= T$) but was discharged in the next period t then the actual LOS is $T + t$.

The probability, p, that the patient is discharged for the time LOS $= T + t$ is

$$p = \int_0^{T+t} f_{ESI}(T)\mathrm{d}T.$$

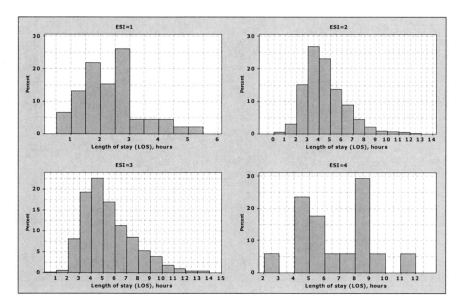

Fig. 3.1 LOS distribution densities for ESI = 1, 2, 3, and 4, accordingly

Based on the rule of total probability, this probability is equal to the probability of discharge at time T and the probability of the opposite event $(1-p)$ times the conditional probability, q, that the patient will be discharged in the following period t, i.e.,

$$\int_0^{T+t} f_{ESI}(T)dT = \int_0^T f_{ESI}(T)dT + \left(1 - \int_0^T f_{ESI}(T)dT\right) \times q_{ESI}.$$

Hence,

$$q_{ESI}(T,t) = \frac{\int_0^{T+t} f_{ESI}(T)dT - \int_0^T f_{ESI}(T)dT}{1 - \int_0^T f_{ESI}(T)dT} = \frac{F_{ESI}(T+t) - F_{ESI}(T)}{1 - F_{ESI}(T)},$$

where $F_{ESI}(T)$ is the cumulative length of stay distribution for each ESI, respectively.

Littig and Isken (2007) used a similar formula (sometimes called the hazard ratio) to compute the conditional probability that the patient will leave in the next time period (24 h) given the current LOS and cumulative LOS distributions (applied to the entire hospital occupancy prediction model).

Examples of LOS distribution densities for four ESI along with their cumulative distributions for a case study hospital are given in Figs. 3.1 and 3.2.

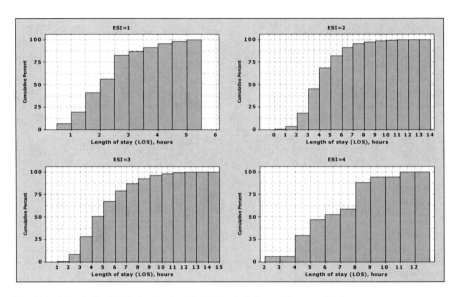

Fig. 3.2 Cumulative LOS distributions for ESI = 1, 2, 3, and 4, accordingly

	INPUT			OUTPUT	
ESI	Number of patients in ED	Current LOS=T hrs	Lead Discharge Time, t hrs	Probability of discharge in the next time period, t hrs	Expected number of patients to be discharged in the next t hrs
ESI=1	5	4	2	0.792	4
	6	5	2	0.480	3
Total ESI=1 in ED	11			Total ESI=1 expected to be discharged in the next 2 hours	7
ESI=2	6	8	2	0.654	4
	7	9.5	2	0.607	4
Total ESI=2 in ED	13			Total ESI=2 expected to be discharged in the next 2 hours	8
ESI=3	5	10	2	0.719	4
	8	6	2	0.595	5
Total ESI=3 in ED	13			Total ESI=3 expected to be discharged in the next 2 hours	9
ESI=4	3	8	2	0.873	3
	4	7	2	0.521	2
Total ESI=4 in ED	7			Total ESI=4 expected to be discharged in the next 2 hours	5

Fig. 3.3 Example of the calculator input and output panels for anticipated number of discharged ED patients

Cumulative distributions for each ESI as functions of LOS were approximated by 3rd order polynomials. These polynomials were substituted in the formula for q_{ESI}, and this formula was used in an Excel spreadsheet calculator.

If the number of patients with the current LOS $= T$ is $N_{ESI}(T)$, then the number of patients expected to be discharged in the following period, t, is $N_{ESI}(T,t)$. It is calculated as

$N_{ESI}(T,t) = N_{ESI}(T) \times q_{ESI}(T,t)$ (rounded to the nearest integer greater than $N_{ESI}(T,t)$).

An example of the calculator input and output is presented in Fig. 3.3. A lead discharge time, t (following the current LOS), was assumed here to be 2 h. Of course, any reasonable values for the number of patients with the corresponding ESI, as well as any current LOS and the lead discharge time, can be plugged in the input panel. The output panel provides the ESI breakdown for the corresponding probabilities of discharge in the next t hours, as well as the expected number of patients to be discharged in the next t hours.

In this example, patient current LOS is from 4 to 10 h (with corresponding breakdown by ESI), respectively. The output panel indicates the number of patients that are expected to be discharged in the next $t = 2$ h from the census checking time (with corresponding breakdown by ESI).

It should be noted that the accuracy of predictions strongly depends on the LOS distribution data for each ESI. Therefore special attention should be paid on representativeness and validity of LOS data. LOS data should be collected for a long enough period of time (about a year or so) including possible seasonal variations.

Chapter 4
Forecasting Time Series

Abstract This chapter includes two problems for forecasting of a time series using past data points. It is argued that the past data points used for forecasting of the future data points should be strongly correlated with each other. It is illustrated that the strongly correlated past data points can be identified from the autocorrelation function of the time series. It is further illustrated that a powerful forecasting procedure for the time series can be a recursive technique. Its application is demonstrated using, as examples, annual patient volume forecasting, as well as forecasting of the seasonal variation of the hemoglobin A1C level.

Keywords Time series • Patient volume forecasting • Autocorrelation function • Time-lag • Recursive forecasting • Seasonal variation

4.1 Forecasting Patient Volumes Using Time Series Data Analysis

Forecasting the future patient volume (demand) is an important step in managerial decision-making, such as planning facility capacity and properly budgeting and allocating resources needed to meet the demand.

As an example, let us consider a problem of forecasting the future patient volumes for a facility using its annual patient volumes for a number of previous years.

The annual patient volumes for 13 years, from 1997 to 2009, are presented in Table 4.1.

For planning budgets and staffing, the facility management needs to forecast annual patient volumes for the future years, from 2010 to 2015.

A. Kolker, *Healthcare Management Engineering: What Does This Fancy Term Really Mean?*, SpringerBriefs in Health Care Management and Economics, DOI 10.1007/978-1-4614-2068-2_4, © Alexander Kolker 2012

Table 4.1 Annual patient volumes

Year	Patient volume
1997	9,400
1998	9,100
1999	9,966
2000	8,900
2001	10,052
2002	9,700
2003	11,200
2004	10,090
2005	12,772
2006	13,130
2007	16,867
2008	17,725
2009	18,225

4.1.1   Traditional Management Approach

The simplest method to forecast the annual patient volume trend (or some workload volume tend) is to assume some percent of annual growth from the last year on the record, say 2% or 3%. Using this assumption, the calculations are simple and some-times presented as plausible "scenarios" for different growth percents. However, such scenarios are actually a wild guess; they assume unlimited growth and provide no insights into trends or the pattern in the data.

Some simple enough smoothing statistical methods are recommended to use, such as polynomial regression for extrapolation (linear and nonlinear), Box-Jenkins, exponential smoothing (single and double), Holt-Winters, weighted moving averages, autoregressive integrated moving averages (ARIMA), and some others (Ozcan 2009; Langabeer 2007).

Smoothing statistical forecasting methods are based on the idea of finding a pattern in the time series data, and then extrapolating this pattern to the future in order to generate forecast.

However, before applying any forecasting method, the number of previous data points that have to be used to make the future predictions should be identified.

4.1.2   The Number of Past Data Points that Have to be Used for Making a Forecast

It seems reasonable to assume that too "old" data points do not practically affect the most recent data points, let alone the future data. For example, it is unlikely that the 13-years-old patient volume data from 1997 would affect the 2009 patient volume, and beyond. On the other hand, it is likely that the 2008 patient volume is more closely related (i.e., correlated to some extent) to the 2009 patient volume.

Therefore, it is reasonable to expect that the 2008 data point will affect to some extent the future data points (forecasts).

The maximum number of "steps back to the past", at which the older data points are still strongly correlated to the newer data-points (the correlation cutoff time lag k), can be estimated using an autocorrelation function (ACF) of a time series. The data points that are strongly correlated to the newer ones can be included for making the forecast. The data points that are weakly correlated to the newer ones (or not correlated at all) should not be included for forecasting; otherwise, the forecast will likely be skewed.

Abraham et al. (2007) point out that models with more past data points (higher order models) may lead to lower forecast errors. At the same time, including too many past data points results also in an increased risk of over-fitting (equivalent to fitting too many parameters in a regression model). On the other hand, these authors argue that the large number of past data used for estimating the model coefficients produce more stable estimates. Therefore they recommend using all available past data points for estimating model coefficients. This is true if the goal is limited to accurate fitting of available data points. However, the actual goal is forecasting the future data beyond available data points. Therefore the use of all available past data points to fit model coefficients (even those data points that are too "old" and weakly correlated with the "newer" ones) will likely result in skewed coefficients and, consequently, skewed forecast.

An autocorrelation function of a time series is a measure of a linear interdependency between the data points separated by k time units (time lag). For a discrete time series of length n with the data points y_i ($i = 1, 2, \ldots, n$), the normalized unbiased ACF for each time lag k ($k = 0, 1, 2, \ldots, m < n$) is calculated as

$$\text{ACF}(k) = \frac{1}{(n-k)\sigma_1(k)\sigma_2(k)} \sum_{i=1}^{n-k} (y_i - m_1(k)) \times (y_{i+k} - m_2(k)),$$

where $m_1(k)$ and $\sigma_1(k)$ are the average and the standard deviation of the first $n - k$ data points, i.e.,

$$m_1(k) = \frac{1}{n-k} \sum_{i=1}^{n-k} y_i, \ \sigma_1^2(k) = \frac{1}{n-k} \sum_{i=1}^{n-k} (y_i - m_1(k))^2.$$

The $m_2(k)$ and $\sigma_2(k)$ are the average and the standard deviation of the last $n - k$ data points

$$m_2(k) = \frac{1}{n-k} \sum_{i=k+1}^{n} y_i, \ \sigma_2^2(k) = \frac{1}{n-k} \sum_{i=k+1}^{n} (y_i - m_2(k))^2.$$

ACF(k) as a function of time lag k goes from 1 (at $k = 0$) to close to 0 for larger k-values, usually (but not always) with some oscillations around the time lag axis k with the decreasing amplitude.

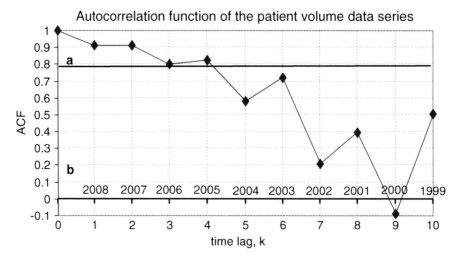

Fig. 4.1 Autocorrelation function of the patient volume time series. Panel (**a**)—ACF(*k*) statistically different from zero. Panel (**b**)—ACF(*k*) are not statistically different from zero at 0.05 significance level

A possible measure of a correlation cutoff value k is the first zero-crossing of the time lag axis (if the crossing occurs). Sometimes, the correlation cutoff lag is defined as the smallest value k that makes $\text{ACF}(k) < K$, where $K = 0.5$ or $K = e^{-1} = 0.37$ (Schuster 1998).

However, it is more justified to compare the computed ACF(*k*) with the critical level required to reject the null hypothesis that ACF(*k*) at the specific lag k was generated by a completely random time series, i.e., it is statistically indistinguishable from zero at some level of significance.

It turns out that the statistic $t = \text{ACF}(k) \times \sqrt{\dfrac{n-k-2}{1-\text{ACF}(k)^2}}$ is distributed as

t-distribution with $n-k-2$ degrees of freedom (Press et al. 1988). Thus, the correlation cutoff lag is the maximum value of k at which the above t-statistic becomes less than the critical t-distribution value at, say, 95% confidence level (at the α level of significance 0.05).

The autocorrelation function ACF(*k*) of the patient volume time series from Table 4.1 is presented in Fig. 4.1.

The ACF(*k*) values are statistically different from zero according to the above statistic for the values $k = 1$ to $k = 4$. Hence, the data points that are strongly correlated in this case go back to the past for about 5 years (ACF(0) = 1 at 2009 corresponds to $k = 0$, i.e., no time lag; each consecutive year corresponds to one time lag unit, i.e., $k = 1, 2, …, 10$). Thus, only the most recent five patient volume data points from 2005 to 2009 with strong interdependency (linear correlation larger than about 0.8) should be used to generate the forecast for the future years.

Table 4.2 Prediction capability of some traditional smoothing forecasting models

Forecasting model	Forecasted value for 2009 using the data points for 2005–2008	Forecasting error (%)
Growth curve	20,369	11.8
Single exponential	15,427	15.3
S-curve trend	17,880	1.9
Winters'	20,449	12.2
Polynomial quadratic	20,397	11.9
Moving average	17,296	5
ARIMA	16,903	7.2

4.1.3 Validation of Some Typical Forecasting Models

In order to check how good some forecast generating models are, the patient volume data points for the years 2005 to 2008 from Table 4.1 were used to predict the data point for the 2009 year. This data point was compared with its known actual value, 18,225. The forecasts, presented in Table 4.2, were generated using the statistical software package, Minitab 15 (with the default software parameters, where applicable).

As can be seen from Table 4.2, none of these traditional models are particularly good at predicting even 1 year ahead. Therefore, it is unlikely that they will provide a reasonably reliable forecast for a longer period of time. Examples of application of four forecasting models for 6 consecutive years from 2010 to 2015 using the last five data points from Table 4.1 are presented in Fig. 4.2.

Regression polynomials (2nd and 3rd order) resulted in an apparent strong downward trend after the year 2011. These models are not shown in Fig. 4.2 due to limited space, as well as Moving Average and ARIMA models.

None of the longer term forecasts generated by these models seem reliable. Indeed, the growth curve and Winters' models forecast that by the end of 2015 the patient volume will be about 35,000 or 30,500, accordingly. These forecasts seem way too large and unlikely.

On the other hand, the single exponential model predicts the constant patient volume at the 18,500 level for all years from 2010 to 2015, while S-curve model predicts a very slow growth with asymptotic constant value at the 18,847 level. These longer term forecasts are also seem unreliable because they indicate practically no change in patient volume at all for the years ahead.

4.1.4 Management Engineering Approach

Since traditional forecasting methods are apparently not adequate enough in this particular situation, another technique can be used. This technique is based on the theory of digital recursive linear filters that are widely used in digital signal processing applications (Press et al. 1988).

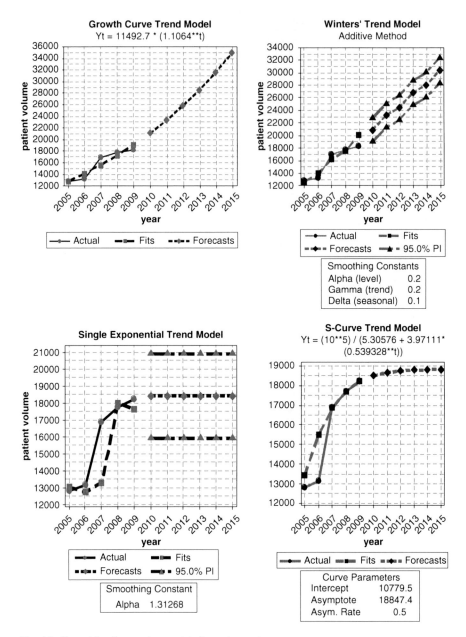

Fig. 4.2 Smoothing forecasting models for patient volume

Let y_i ($i = 1, 2, \ldots, n$) are the data points that are equally spaced along a time line, and one wants to use n consecutive values of y_i to predict the $n + 1$ data point. The equation for predicting the next data point of a time series from the previous values is

$$y_n = \sum_{j=1}^{m} d_j y_{n-j} + x_n,$$

where d_j is a set of prediction coefficients, m is the number of coefficients (model's order), and x_n is the discrepancy of the prediction at time-step n and the true value, y_n.

Prediction coefficients, d_j, are calculated in a way that minimizes the discrepancy, i.e., makes $|x_n| \ll |y_n|$, or $\sum_n |x_n|^2 \ll \sum_n |y_n|^2$. Thus, using a right set of coefficients d_j, one can predict the future data points of a time series from a record of its past.

Because the predicted future data points, y_n, are generated using previously calculated data points, y_{n-j}, such a procedure is called the *recursive* procedure; its predicting behavior quickly becomes much more complex than a straight line, or a polynomial. It is especially successful at predicting time series that are rather smooth and oscillatory, though not necessarily periodic (Press et al. 1988).

However, in order to achieve its full usefulness, recursive procedure must be stable. Recursive procedures feed on their own output; therefore they are not always stable, i.e., some particular "bad" sets of coefficients, d_j, can generate an exponentially growing output.

The condition that the recursive prediction procedure is stable is that all n complex roots of the characteristic polynomial equation

$$z^n - \sum_{j=1}^{n} d_j z^{n-j} = 0$$

are inside the unit circle, i.e., satisfy the condition $|z| \leq 1$. Press et al. (1988) provide a detailed description of the computational procedure for using recursive prediction of the future data points of a time series using the above equations.

Similarly to the traditional smoothing forecasting models illustrated above, the recursive procedure was tested using the previous four patient volume data points for the years 2005 to 2008 in order to predict the fifth data point for the year 2009, which was compared with the known actual value. Forecasted 2009 patient volume was 18,114, while the actual value for 2009 was 18,225. The prediction error in this case was rather small, about 0.6%. Thus, this forecast is much better than that provided by smoothing forecasting methods.

Of course, one should not always expect such a good predicting accuracy. However, Press et al. (1988) pointed out that in many situations the recursive

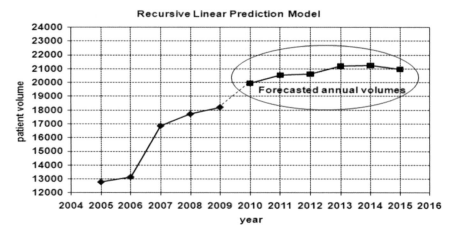

Fig. 4.3 Patient volume forecasting based on recursive linear prediction model

forecasting turns out to be vastly more powerful than any kind of a smoothing or polynomial extrapolation used in a traditional forecasting.

Forecasted values for 6 consecutive years from 2010 to 2015 using the most recent five data points from Table 4.1 are presented in Fig. 4.3.

It is seen that the recursive linear forecasting is more reasonable than some traditional smoothing forecasting models presented in Fig. 4.2. Indeed, recursive forecasting predicts neither an unlimited growth in patient volume over the years, nor its constant value. Instead, the forecast indicates some growth in patient volume up to the year 2013, flattening to 2014 and then a slight decline from 2014 to 2015.

Of course, no forecast can be very accurate; forecasting errors are inevitable. In general, it is impossible to accurately predict the future based on the past.

Nonetheless, if the underlying factors are stable enough for the forecasting time horizon, and the most recent data points from the past are used that are strongly correlated to each other, then a recursive forecasting is capable of providing a reliable enough forecast that can be valuable for efficient managerial decision-making.

4.2   Forecasting Time Series with Seasonal Variation

Time series to be forecasted often include seasonal variability and/or some other periodic or nonperiodic variations. Many physiologic processes have been reported to vary seasonally in both healthy people and people with chronic diseases. They include blood pressure, heart rate, lipid level, etc. Cardiovascular events, strokes, and mortality have a distinct seasonal fluctuation. Tseng et al. (2005) reported a significant seasonal variation in population monthly hemoglobin A1C level over

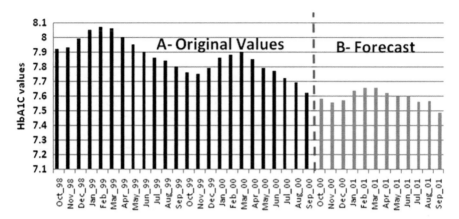

Fig. 4.4 Monthly mean hemoglobin A1C values for US veterans with diabetes. Two years period: October, 1998 to September, 2000. (**a**) Original values (Tseng et al. 2005). (**b**) 12 months forecast for 2001 based on recursive linear prediction model

2 years (from October 1998 to September 2000) among US diabetic veterans. A1C values were higher in winter (cold months) and lower in summer (warmer months). Original monthly mean A1C data are presented in Fig. 4.4 (left panel—a).

It is clearly seen that the original 2-year data are seasonably variable with a downward trend.

In order to predict (forecast) A1C monthly values with seasonal variability for the next year 12 months several techniques could be used.

4.2.1   Traditional Management Approach

Typically, if a time series contains seasonal variation combined with a trend the time series is decomposed on the pure trend and seasonal (periodic) variation. Removing seasonality is usually done by dividing each data point by its seasonal index. In order to calculate a seasonal index the following procedure can be used: (1) the data smoothed using a centered moving average with a length equal to the length of the seasonal cycle (if the seasonal cycle length is an even number, a two-step moving average is required to synchronize the moving average correctly); (2) the moving average is either divided into (multiplicative model) or subtracted from (additive model) the data to obtain what is often referred to as raw seasonal values; (3) for corresponding time periods in the seasonal cycles the median of the raw seasonal values are determined. For example, if 24 consecutive months of data (2 years) are available then the median of the 2 raw seasonal values are determined corresponding to January, to February, and so on; (4) the medians of the raw seasonal values are adjusted so that their average is one (multiplicative model) or zero (additive model). These adjusted medians constitute

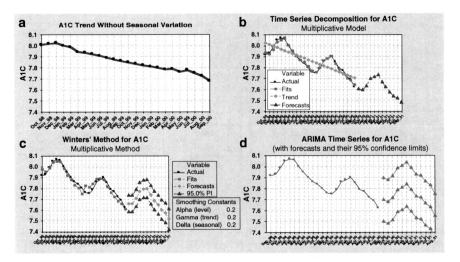

Fig. 4.5 Twelve-month forecasting of 2-year A1C data with seasonal variation: (**a**) Trend of original A1C data without seasonal variation; (**b**) forecasting using decomposition on trend and seasonal variation using seasonality indices; (**c**) forecasting using Winters' method; (**d**) forecasting using ARIMA

the seasonal indices; (5) the seasonal indices are used to seasonally adjust the data; and (6) a trend line is fitted to the seasonally adjusted data using least squares regression.

The data can be detrended by either dividing the data by the trend component (multiplicative model) or subtracting the trend component from the data (additive model). Forecasting is done on the deseasoned (seasonally adjusted) trend (typically linear trend). Seasonal variation is then returned back to the forecast using the seasonal indices calculated earlier. This procedure can easily be done using time series analysis option in statistical packages, such as Minitab or SPSS.

Application of this method for 2-year A1C data is presented in Fig. 4.5.

Seasonally adjusted data (pure trend with the removed seasonal variation) is presented on panel A. This is a downward trend and it is very close to the linear. Twelve-month forecast based on this trend and reinstated seasonal variation is given on panel B. Notice that the amplitude of the seasonal variation remains the same for all years because the same seasonal indices are used.

Forecasting using Winters' method is presented on panel C. Winters' method smoothes data by Holt-Winters exponential smoothing. In contrast to the decomposition method, Winters' method calculates dynamic estimates for three components: level, trend, and seasonal. Dynamic method analyzes patterns that change over time and estimates are updated using neighboring values.

Application of ARIMA method is given in panel D. ARIMA stands for Autoregressive Integrated Moving Average. The terms in the name, i.e., Autoregressive, Integrated, and Moving Average represent filtering steps taken in

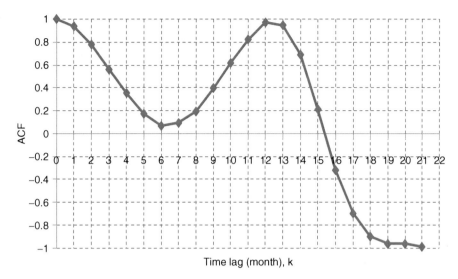

Fig. 4.6 Autocorrelation function (ACF) of the A1C time series with seasonal variation

constructing the ARIMA model until only random noise remains. The forecast profile depends upon the model that is fit. The advantage of ARIMA modeling compared to the simple forecasting and smoothing methods is that it is more flexible in fitting the data. However, identifying and fitting a model may be time-consuming, and ARIMA modeling is not easily automated.

These three methods generally provide similar forecasts with seasonal variation, and it is not easy to pick the best in this case.

4.2.2 Management Engineering Approach

It was mentioned in Sect. 4.1 that recursive linear forecasting is a powerful method especially successful at forecasting time series that are rather smooth and oscillatory, though not necessarily periodic. The 2-year A1C data represent such a smooth and oscillatory time series. Therefore it makes sense to apply this method and compare it with more traditional methods considered earlier. Recursive linear forecasting does not require any decomposition of a time series into its trend and seasonal variability. It can be directly applied to the original data to generate the forecast. However, as in the previous section, the number of past data points needed for forecasting should be determined. This can be done using the autocorrelation function (ACF) of a time series.

The ACF of the A1C time series calculated using the formulas from Sect. 4.1 is presented in Fig. 4.6.

It follows from the t-statistics that the data points for time lags $k = 4, 5, 6, 7, 8$ as well as for $k = 15$ and 16 are not statistically significant at the 0.05 level. Formally, it means that these data points should not be used for generating a forecast, as discussed in Sect. 4.1. However some older data points, such as $k = 10$–14 and $k = 17$–21, are significant and, hence, should be used for forecasting. The reason for the situation when the older data points seem stronger associated (correlated) than the newer data points is a seasonal variability. Therefore all data points up to $k = 21$ (at least 22 months out of total 24 months) should be used for forecast-generating in this case. ($ACF(k)$ and cutoff t-statistics for a last few data points $k = 22$–24 cannot be calculated with a reasonable accuracy).

In order to demonstrate that practically all data points are actually strongly correlated in this case the pure trend with the removed seasonal variation shown in Fig. 4.5 (panel A) can be used to calculate $ACF(k)$. It turns out that $ACF(k)$ for pure trend without seasonal variation is indeed statistically significant at the 0.05 level up to the time lag $k = 21$, i.e., 22 months data points (the $ACF(k)$ plot for the deseasoned trend is not shown here to save space). Thus, 22 data points out of total 24 data points were used to generate 12-month A1C forecast. This forecast is shown in Fig. 4.4 (panel B).

Recursive linear forecast method clearly reflects the seasonal variability with maximal values in February and March and minimal values in September and October. However, the amplitude of the seasonal variability for forecasted data points is smaller than that for traditional methods. This is because in traditional methods the amplitude of the original seasonal variability is retained, and it is simply overlaid on the pure forecasted trend (usually linear). In contrast, in recursive forecasting the amplitude of the seasonal variability is not necessarily retained in full, but it is directly forecasted from the available data pattern.

Chapter 5
Business Intelligence and Data Mining

Abstract This chapter includes an application of business intelligence (BI) and data mining (DM) based on advanced multivariate data analysis. Two examples are presented: principal component decomposition of the large dataset of original observational data to identify contributing independent variables (BI), and cluster analysis (DM) to identify the distinct groups of population ZIP codes (clusters) in terms of patient contribution margins.

Keywords Business intelligence • Principal components • Observational database • Correlation • Linear combination • Regression analysis with principal components • Data mining • Hierarchical cluster • Dendrogram • k-means cluster • Linkage • Similarity

Business Intelligence (BI) is loosely defined as a combination of quantitative methods that help to extract meaningful information or pattern from large raw databases for managerial decision-making and developing current or future business actions. Data mining can be defined as the analysis of observational datasets to find unsuspected relationships and to summarize the data in novel ways that are useful to the data owner (Hand et al. 2001).

The problems in this section illustrate an area of application for management engineering that is largely overlapped with some applications of business intelligence such as business analytics and data mining.

The first problem is an advanced statistical multivariate data analysis methodology for identifying a few most important contributing factors from a large raw database (Sect. 5.1). The second problem is an application of cluster analysis to identify the groups of objects (clusters) in such a way that the objects within one cluster have much higher similarity compared to objects within another cluster (Sect. 5.2).

A. Kolker, *Healthcare Management Engineering: What Does This Fancy Term Really Mean?*, SpringerBriefs in Health Care Management and Economics, DOI 10.1007/978-1-4614-2068-2_5, © Alexander Kolker 2012

5.1   Multivariate Database Analysis: What Population Demographic Factors Are the Biggest Contributors to Hospital Contribution Margin?

A hospital plans a major market share expansion to improve its long-term financial viability. The management would like to know what population demographic factors specific to the ten local core service area zip codes are the most important contributors to the financial contribution margin (CM $). Contribution margin is defined as the difference between payments collected from a patient and the patient variable costs.

A set of population demographic data was collected for ten local area zip codes as well as corresponding median contribution margins from each zip code (CM $).

The following groups of demographic variables were collected for each zip code as the percentage of the total zip code population (actual data are not shown here because of limited space):

- Eight Age categories: 18–24, 25–34, 35–44, 45–54, 55–59, 60–64, 65–74, and 75+ years old
- Nine Educational categories: less than high school (less HS), some HS, high school (HS), some College, associate degree (AD), bachelor degree (BD), master degree (MD), professional degree (ProD), and PhD
- Ten Income categories (annual): less than $15K, $15K–$25K , $25K–$35K, $35K–$50K, $50K–$75K, $75K–$100K, $100K–$150K, $150K–$250K, $250K–$500K, and $500K+
- Five occupational categories: Health care, Labor, Professional/Administrative, Public Service, and Service industry

Thus, there are total 32 data variables for each zip code included in the database.

5.1.1   Traditional Management Approach

Because of the large number of variables, it is difficult to select the most important factors for the zip codes with the largest CM $. For example, it could be expected that older and more affluent patients (e.g., 75+ years old and in annual income category $100K+) contribute more to CM $ than younger and/or lower income patients. However, actual data indicate that the zip codes with the largest CM $ have lower percentage of the above categories than the other zip codes with lower CM $.

On top of that, many of these categories are highly correlated. This means that there is a redundant information and uncertainty in the data that makes it difficult to attribute the contribution of one or more variables to CM $.

For example, it is reasonable to expect a high positive correlation of low education and low income, and a negative correlation of high education degrees (say, MD or PhD) and low income. Indeed, using demographic data, it was calculated that a correlation coefficient of the categories "some HS" and "annual income less than $15K" is 0.899, while the same income and "master degree" (MD) are correlated negatively at −0.736. On the other hand, MD and "$150K–$250K" income category have a high positive correlation coefficient 0.873. This means that it is very likely that those with "some HS" earn annually less than $15K, while there is a strong tendency that those with MD educational level earn annually in the range $150K–$250K.

Obviously, performing such a paired correlation analysis for all 32 variables (496 pairs!) for each of ten zip codes is impractical. Besides, knowing the linear correlation coefficient does not help to reduce redundant information and extract meaningful information for separate contributing factors.

An attempt to perform a linear regression analysis using all 32 variables as predictors and CM $ as a response function results in a regression equation with low goodness of fit (R-sq(adj) is 8.6%), and with all coefficients that are not statistically significant at the 5% significance level (p-values is in the range from 0.1 to 0.97) with the huge variance inflation factors (VIF) that are in the range from a few dozen to a few millions. This is an obvious indication of a serious multicollinearity problem in the original dataset that leads to the failure of the traditional regression analysis to extract meaningful information for contributing factors (variables).

Therefore, an alternative technique should be employed. One of the most powerful methodologies for multivariate data analysis is based on principal component decomposition of the original database matrix (Jobson 1992; Glantz and Slinker 2001).

5.1.2 Management Engineering Approach

Principal component decomposition methodology allows one to perform a multivariate correlation analysis and identifies redundant variables that carry little or no independent information while retaining only a few mutually uncorrelated principal variables (components) that contain practically all original information. This technique is a special case of a matrix approximation procedure called singular value decomposition. The higher the level of correlation between the columns of data of the original matrix, the fewer the number of new (principal) variables is required to describe the original dataset.

More formally, given an original data matrix $\mathbf{X}$ representing n observations on each of p variables $X_1, X_2, ..., X_p$ the purpose of principal component analysis is to determine r new variables PC_r (PC—principal component) that can be used to best approximate variation in the p original X variables such as linear combinations

$$PC_1 = v_{11}X_1 + v_{21}X_2 + ... + v_{p1}X_p$$

$$\cdots\cdots\cdots\cdots$$

$$PC_r = v_{1r}X_1 + v_{2r}X_2 + ... + v_{pr}X_p$$

Table 5.1 Eigen value analysis of the demographic data correlation matrix

Eigen value	18.537	4.801	3.433	1.970	1.351	0.690	0.590	0.395	0.234
Proportion	0.579	0.150	0.107	0.062	0.042	0.022	0.018	0.012	0.007
Cumulative	0.579	0.729	0.837	0.898	0.940	0.962	0.980	0.993	1.000

(Best approximation is calculated as the minimized sum of squared deviations of the PC approximations and the variables in original data matrix **X**).

The solutions to the problem are the eigenvalues, λ_j, $j = 1, 2,...,s$, and the corresponding eigenvectors, v_j, $j = 1,2,...,s$. The elements of the jth eigenvector define the jth PC associated with the data. The jth eigenvalue is a measure of how much information is retained by the jth PC. A large value of λ_j (compared to 1) means that there is a substantial amount of information retained by the corresponding jth PC, whereas a small value means that there is little information retained by jth PC.

All eigenvalues must add up to the total number of the original independent variables, p, i.e., $\sum_{j=1}^{s} \lambda_j = p$. Thus, if some eigenvalues are large, then the others should be small. This illustrates the principle of information conservation: the total amount of information in the original dataset is not changed because of PC decomposition; rather, it is rearranged in the form of a number of linear combinations of the original variables in such a way that main information holders (linear combinations – PCs) are clearly identified, significantly reducing thereby the number of independent variables that retain the same amount of information.

For this particular problem, the principal component analysis of the demographic matrix has been carried out using the Minitab statistical software package (version 15). Results are presented in Table 5.1.

Thus, only nine principal components (nine linear combinations of the original variables) are required to account for all 32 original variables. Only five principal components are enough to approximate 94% of the original data. This indicates that a lot of variables in the original data matrix are indeed highly correlated and contain no new information; most of them form a so-called information noise that hampers extracting meaningful information for contributing factors.

Notice also that the sum of all nine eigenvalues in Table 5.1 is equal to the number of the original data variables, i.e., 32, as it is required by the correct PC decomposition technique.

The next step is to perform regression analysis in order to relate CM $ (response function) to mutually uncorrelated PC variables, and then to determine the original variables (factors) that contribute the most to CM $.

Regression analysis with PC has substantial advantages over regular regression analysis with the original data, and it is much more reliable (Jobson 1992; Glantz and Slinker 2001).

Because PCs are mutually uncorrelated, the variation of the dependent variable (CM $) is accounted for by each component independently of other components, and their contribution is directly defined by the coefficients of the regression

Table 5.2 Best subsets regression with PCs

Variables	R-sq (adj)	Mallow Cp	PC1	PC2	PC3	PC4	PC5	PC6	PC7	PC8
3	50.9	6.5				X		X		X
3	49.9	6.6			X	X		X		
4	60.7	5.7			X	X		X		X
4	59.4	5.8	X			X		X		X
5	73.7	5.0	X		X	X		X		X
5	62.5	6.3		X	X	X		X		X
6	80.4	5.7	X	X	X	X		X		X
6	72.2	6.4	X		X	X		X	X	X
7	**81.6**	**7.1**	**X**	**X**	**X**	**X**		**X**	**X**	**X**
7	71.6	7.6	X	X	X	X	X	X		X
8	65.3	9.0	X	X	X	X	X	X	X	X

equation. Since PCs are mutually uncorrelated, the presence of any one PC does not affect the regression coefficients of the other PC.

Usually, if the purpose of PC analysis is reduction of the number of independent variables and the search for the underlying data pattern, the PCs with the eigenvalues close to zero (minor PCs) can be dropped because they carry on little or no information. However, in multiple regression with PC it is not a good idea to drop the minor PCs because doing so can introduce an uncertain bias in the regression coefficients (Jobson 1992; Glantz and Slinker 2001). In this particular case, the number of PCs that account for all original data is relatively small (nine). Therefore, all nine PCs have been retained for performing multiple regressions.

In order to perform the regression, we first find the best subset using PCs as independent variables. Results for the predictors PC1 to PC8 are presented in Table 5.2 (somewhat shortened to save space).

The best subset regression identifies the best-fitting regression models that can be constructed with as few predictor variables as possible. All possible subsets of the predictors are examined, beginning with all models containing one predictor, and then all models containing two predictors, and so on. The two best models for each number of predictors are displayed.

Each line of the Table 5.2 represents a different model. An X symbol indicates predictors that are present in the model.

In this case, the best subset is the first line with seven variables (highlighted bold) because it has the largest R-sq(adj) and the Mallow Cp coefficient that is reasonably close to the number of predictors. This subset includes seven predictors PC1 to PC4 and PC6 to PC8.

A subset analysis with the predictors PC2 to PC9 (not shown here) results in overall lower R-sq(adj), and it is not used for further regression. This is expected because PC1 with the largest eigenvalue 18.537 is not included. (Notice that the use of all nine PCs at once to find the best subset is not possible because the total number of regression coefficients and constant should be less than the number of observations, ten in this case).

Table 5.3 Eigenvector coefficients for statistically significant principle component PC6

Variable	PC6	Variable	PC6
Age 18–24	−0.159	Some HS	−0.185
Age 25–34	−0.108	Income <$15K	−0.18
Age 35–44	0.132	Income $15K–$25K	−0.077
Age 45–54	−0.046	Income $25–$35K	0.393
Age 55–59	−0.039	Income $35K–$50K	0.254
Age 60–64	−0.133	Income $50K–$75K	−0.097
Age 65–74	−0.053	Income $75K–$100K	−0.078
Age 75+	0.193	Income $100–$150K	−0.084
AD degree	0.146	Income $150K–$250K	0.182
BD degree	0.084	Income $250–$500K	0.057
PhD	0.103	Income $500K+	−0.256
HS	−0.111	Occupation: Health care	−0.03
Less than HS	0.095	Labor	0.014
MD	0.147	Professional/administrative	0.166
Pro degree	0.221	Public service	−0.359
Some college	−0.081	Service industry	−0.417

Regression equation with subset predictors PC1 to PC4 and PC6 to PC8 is

$$CM \ \$ = 13,907 - 115 \ PCI - 162 \ PC2 - 278 \ PC3 - 483 \ PC4$$
$$- 1143 \ PC6 + 319 \ PC7 - 846 \ PC8$$

with R-sq $= 95.9\%$ and R-sq (adj) $= 81.6\%$. Thus, this equation is rather accurate in what it accounts for most of the response function (CM $) variability. Variable inflation factor (VIF) for all predictors is exactly 1 versus huge VIF for the regression equation with the original 32 variables. This confirms that all predictors are mutually uncorrelated and there is no multicollinearity issue at all.

At the same time, only one predictor, PC6, turns out to be statistically significant at the 5% significance level: PC6 (p-value $= 0.05$). Because PC predictors are orthogonal (mutually uncorrelated), all terms that are not statistically significant can be removed from the regression equation without affecting the remaining coefficients. Nonetheless, one principal component, PC6, retains all 32 original variables.

Because PC6 regression coefficient is negative, relatively large negative PC6 eigenvector coefficients make largest contributions into of CM $. The eigenvector coefficients for PC6 are presented in Table 5.3. These coefficients represent the weight of each variable into PC6.

Assuming that large negative coefficients are those greater than 0.1 in absolute values, an examination of the eigenvector coefficients from this table results in the following primary contributing variables (factors) to CM $:

Age category: "18–24 years old" followed by "60–64 years old"
Education category: "some HS" followed by "HS"
Income category: "$500K +" followed by "less than $15K"
Occupation category: "service industry" followed by "public service"

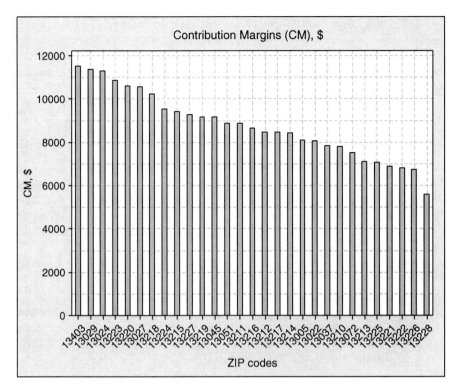

Fig. 5.1 Median contribution margins (CM, $) for some zip codes

It is worth noting, for example, that such contributing factors as "some HS" and "less than $15K" are highly positively correlated. Therefore, it is likely that they include practically the same population with high poverty and low education. Most of this population is likely on the government assistance program, and are frequent users of hospital services.

There can be some other possible demographic and societal explanations of results of this analysis, but they are beyond the scope of this book.

Thus, in order to increase the contribution margin, hospital management should focus the marketing campaign on attracting more patients from the zip codes with a higher level of the above primary contributing factors.

5.2   Cluster Analysis: Which Zip Codes Form Distinct Contribution Margin Groups?

A regional community hospital collected contribution margin (CM) data for all patients from local area zip codes. Median CM (in descending order) are shown on Fig. 5.1 (actual zip codes are replaced by dummy zip codes; for Figure clarity, only a limited number of zip codes is presented).

The hospital management wants to get answered the following question using these data: is it possible to combine some zip codes into a few relatively homogeneous groups (clusters) with similar CM in order to analyze which factors make one group (cluster) different from others? Or each of a few dozen zip codes should be analyzed separately to find out why its CM is different from others?

5.2.1 Traditional Management Approach

From a traditional standpoint, statistical hypothesis testing can be used to find out whether the evidence exists of the statistically significant difference between CM of the different zip codes. If there is no statistically significant difference between CM for some zip codes, then these zip codes could be combined into one group; hence, they could be treated and analyzed as one group.

A frequently used statistical test is ANOVA (Analysis of Variance). Using ANOVA, several combinations of zip codes could be tested. It could be concluded, for example, that there was no statistically significant difference in CM at the 0.05 significance level within the group of five top zip codes 13403 to 13220, as well as within the group of six bottom zip codes, 13213 to 13228 (Fig. 5.1). However, there was statistically significantly difference between these groups. Hence, it could be concluded that the data contain at least two distinct groups of zip codes.

However, such an approach does not make any statistical sense. All statistical tests are developed and are applied only to random samples to make some inferences about the population from which these samples were drawn. In this case, CM data are not random samples; they are the entire population data, i.e., CM of all patients seen at the hospital from each zip code. The mean, median, the variance, or any other statistics can, of course, be calculated for the entire population, but no statistical tests are needed and valid for data that describe this entire population.

Therefore, another approach for grouping population data is needed that is not based on random sample statistical testing. Such an approach can be based on data mining methodology using, for example, cluster analysis.

5.2.2 Management Engineering Approach

There are three main data mining methods that are widely used: classification, clustering, and association (Hand et al. 2001). In this section, we focus only on clustering, which is normally performed when little or no information about data structure and groups of data is available.

Both principal component decomposition and cluster analysis are techniques for data volume reduction. Principal component decomposition is used to reduce the number of variables (the number of columns) in the data matrix (for example, it was illustrated in Sect. 5.1 that the number of variables could be reduced from 32 to 9).

Cluster analysis is used to reduce the number of observations of the variables (the number of rows in this data matrix).

The main concept of cluster analysis is proximity between two groups of objects (Jobson 1992). A variety of approaches to the measurement of proximity are available. Among them: (1) single linkage or nearest object-to-object distance; (2) complete linkage of furthest object-to-object distance. This is the opposite of the single linkage measure; (3) the average linkage that is given by the averaging all paired distances of all objects between the groups of objects; and (4) centroid linkage that is defined as the distance between the centers of weight of the groups; and some others (Jobson 1992).

There are several distant-type measures, such as Euclidian and weighted Euclidian; Mahalanobis that takes into account the covariances among the objects; Manhattan or city block metric that is based on the absolute values of the difference between the objects' coordinates; Minkowski metrics; and some others (Jobson 1992).

The average linkage and centroids with Euclidian distance are the most widely used metrics because of their relative insensitivity to extremes or object outliers. However, depending on the type of clusters, other metrics could have advantage.

There are two main clustering procedures: hierarchical and partitional. The clustering procedure that uses an agglomerative hierarchical method begins with all observations being separate, each forming its own cluster. In the first step, the two observations closest together are joined. In the next step, either a third observation joins the first two, or two other observations join together into a different cluster. This process will continue until the final partition is reached. The final partition is defined by the user either by the total number of clusters or by a similarity level. This procedure can be visualized by a so-called dendrogram or tree diagram. The dendrogram shows the manner in which the clusters are formed—either by joining two individual observations, or pairing an individual observation with an existing cluster. It can be seen at what similarity levels the clusters are formed, and the composition of the clusters of the final partition. On the other hand, if the dataset is too large, the dendrogram visualization capability is poor, and it provides little help.

The partitional procedure uses nonhierarchical clustering of observations. K-means clustering, the most widely used partitional method, works best when sufficient information is available to make good starting cluster designation or expected number of clusters, k. K-means procedure first randomly selects k centroids (objects), and then decomposes objects into k disjoint groups by iteratively relocating objects based on the similarity between the centroids and the objects. In K-means, a cluster centroid is the mean value of objects in the cluster. The method is generally more accurate than hierarchical clustering. However, it provides no visualization aid. On the other hand, hierarchical procedure is much slower in computational time than partitional procedure. The later can practically handle much larger datasets than the former.

Cluster analysis of the data presented on Fig. 5.1 was performed with the software package Minitab 15 using hierarchical and K-mean methods with the centroid linkage and Euclidian distance. An example for 5-cluster dendrogram and corresponding cluster partitioning ($k = 5$) are presented in Fig. 5.2.

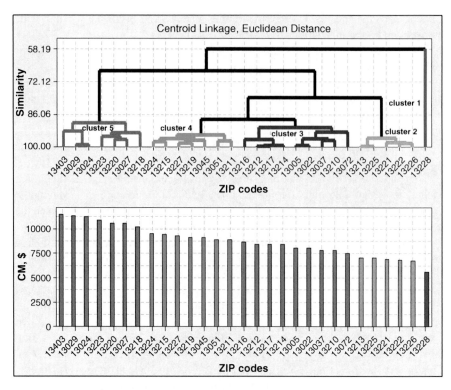

Fig. 5.2 Hierarchical clustering: 5-cluster dendrogram (*top*) and cluster partitioning for CM data from Fig. 5.1 (*bottom*)

It follows from both graphs (top and bottom) that according to hierarchical clustering there is one zip code 13228, which forms 1-object cluster with sharply distinct minimal CM, $5,588 (cluster 1). Five zip codes (13226, 13222, 13221, 13225, and 13213) form cluster 2 with relatively close CM, from about $6,700 to $7,100. Next nine zip codes (from 13072 to 13216) form cluster 3 with the CM range from $7,500 to $8,600. Seven zip codes (from 13211 to 13224) form cluster 4 with the CM range from $8,900 to $9,500, and last six zip codes from 13218 to 13403 form cluster 5 ($10,200 to $11,500).

Comparative results of K-mean clustering procedure for $k=5$ clusters using the same original data are presented on Fig. 5.3.

In contrast to hierarchical procedure, K-mean procedure selects zip code 13403 as 1-object cluster (cluster 1), which has the highest CM, $ 11,518. Next two zip codes (13029 and 13024) form a 2-member cluster 2 with CM from $11,312 to $11,376. 4-member cluster 3 is formed by four zip codes from 13223 to 13027, with CM from $10,873 to $ 10,567, respectively, and so on, as indicated in Fig. 5.3.

It appears that K-mean partitioning in this example is more intuitively appealing and makes more sense than hierarchical partitioning.

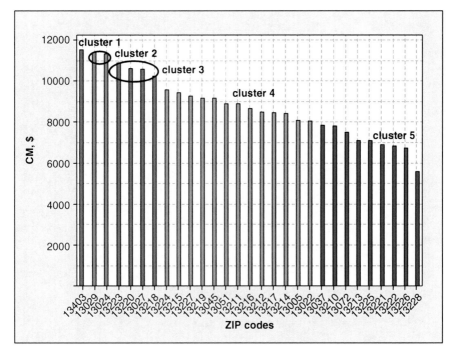

Fig. 5.3 *K*-means clustering: cluster partitioning for CM data from Fig. 5.3

Thus, depending on the required granularity of data, the variable CM with total 29 observations (zip codes) can be reduced to the variable with 5 (or another small number) of observational groups (clusters) with values that are similar to each other within the cluster with a given similarity metric. Thus, if there is some known information available about an object (zip code), one can take full advantage of this information to understand the other objects (other zip codes) in the cluster to which this object belongs. Analysis of a few relatively small groups of similar objects (zip codes) is usually easier than that of the large number of original unclustered observations.

In summary, data mining methods and, specifically, the cluster analysis are now rather widely used in health care for many applications. These applications can include, for example, identifying health insurance fraud, and/or under-diagnosed patients, performing marketing and cost analysis, and many others. On top of that, they can help to obtain frequent patterns from the various databases, and help to recognize relationships among diseases and drugs, and patients' health status (Yoo et al. 2011).

There is a caveat, however, in using cluster analysis that should be kept in mind: there always exists a risk of chance finding a cluster. Regardless of the particular procedure, clustering always produces some result. Even if the data points are randomly and uniformly distributed (totally unstructured), a clustering procedure will suggest existing a group of related data points (clusters) (Peek 2010). Therefore, efforts have been made to develop methods for distinguishing genuine clusters from chance clusters (Tibshirani et al. 2001).

Chapter 6
The Use of Game Theory

Abstract This chapter demonstrates an application of mathematical game theory for allocating cost savings (gains) between cooperating providers. As an alternative to traditional accounting allocation basis, there is a growing interest in cost allocation principles based on logic of game theory. The most widely used method of joint-cost (savings) allocation is the Shapley value, which provides a fair cost allocation based on each participant's estimates of the net benefits (net values) expected from cooperation with other groups of participants (coalitions).

An example is provided for distributing the savings of the bundled payment between four cooperating participants: hospital, physician group, skilled nursing facility, and home health agency.

Keywords Game theory • Shapley value • Coalition • Net benefit • Bundled payment • Savings • Cooperating providers

Hospitals provide numerous examples of conflicts of interests between units/departments and the various stakeholders. It is easy to find many examples of "games" between various "players" in the hospital. For example, there could be a conflict of interests between ED that wants to transfer a patient to ICU or NU quickly, and the ICU that might wish to delay the transfer because it does not have capacity and it is not ready to admit another patient (see problem 2.3.11). Or a nursing unit might be interested to prolong patients' stay to keep its high occupancy level, while the other (noncooperating) units might strive for a quick discharge to increase their turn-around. These and many other types of problems that involve conflicts and decision strategies for cooperative or noncooperative participants (players) present a fruitful ground for application of mathematical game theory.

In the following section, a game theory approach is demonstrated for allocating cost savings between cooperating participants using a widely used concept of Shapley value (Young 1994).

A. Kolker, *Healthcare Management Engineering: What Does This Fancy Term Really Mean?*, SpringerBriefs in Health Care Management and Economics, DOI 10.1007/978-1-4614-2068-2_6, © Alexander Kolker 2012

Table 6.1 Median costs of heart bypass care for individual participating providers

Participant (member)	Hospital {1}	Physician group {2}	SNF {3}	HHA {4}	Total
Median cost for individual participating providers, $	$15,000	$8,000	$6,000	$3,000	$32,000

6.1 Is Distributing of Savings Between Cooperating Providers Fair? The Use of the Shapley Value Concept

Policymakers have grown increasingly frustrated with fee-for-service payment system. Fee-for-service rewards volumes and encourages silos and fragmentation of care. Several provisions of 2010 healthcare reform legislation seek to shift provider payments to value-based approaches that encourage quality improvement and cost reduction. Payment bundling is one such approach.

Payment for all providers of services will be bundled into a single payment that will be paid to a contracting organization. That organization must include a hospital (H), physician group (PG), skilled nursing facility (SNF), and home health agency (HHA). The contracting organization will be responsible for allocation the payments among all providers. Therefore, it must develop a methodology for "fair" distribution of the payments and gain sharing (savings) among providers.

As an example, the median costs of heart bypass surgery and postsurgical care for individual participating providers (a study hospital, physician group, skilled nursing facility, and home health agency) are shown in Table 6.1.

The payments negotiated between the contracting organization and Medicare averaged about 13% lower than the equivalent amounts that would be paid outside of the bundled payment pilot program. This created an immediate cost savings to the Medicare program, but also required the participants to make up the revenue loss in cost savings to break even.

The benefits of increased alignment (cooperation) between hospitals and physicians have been well-established. The bundled payment offers a number of opportunities for hospitals to accrue cost savings through reductions in actual hospital costs (reduction of operating expenses), decreases in length of stay (LOS), reduction or avoidance of readmissions, and management postacute care (Moeller and Evans 2010).

Thus, the contracting organization needs to distribute 13% cost reduction (savings) of the total median cost per case, i.e., $32,000 \times 13\% = \$4,160$ between the cooperating participants: H {1}, PG {2}, SNF {3}, and HHA {4}.

6.1.1 Traditional Management Approach

The most straightforward approach is simply divide the cost reduction (savings) equally between all participants in collaboration, i.e., allocate $4,160/4 = \$1,040$ to each of them. However, this does not seem fair because the role and contribution of each participant to the total gain is different.

Another approach is to share the savings proportionally to the participants' costs, i.e., H {1} savings is $15,000/$32,000×$4,160=$1,950; PG {2} savings is $8,000 /$32,000×$4,160=$1,040; SNF {3} savings is $6,000/$32,000×$4,160=$780; and HHA {4} savings is $3,000/$32,000×$4,160=$390.

However, the savings for smaller (lower cost) participants in this approach might seem too low for them to encourage their active participation in cooperation with other bigger participants.

6.1.2 Management Engineering Approach

By pooling resources (cooperating), the participants usually reduce the total joint costs and realize savings. The question arises is how the realized savings should be shared fairly between the collaborating participants. As elaborated in game theory literature (Young 1994), most proposed fair allocation schemes are based on two main approaches: the Shapley value and the Nucleolus. The latter is an equilibrium that finds the "center of gravity" of the so-called core (if the core exists). The core is defined as a set of inequalities that meet the requirement that no participant or a group of participants pays more than their stand-alone cost.

In this section, only the Shapley value is illustrated because it is easier to calculate and it is more widely used.

The Shapley value is a game theory concept aimed at the "fair" allocation of collectively gained profits (savings) between several collaborative participants (members). The basic idea of the Shapley value is to find the relative (net) contribution of every participant to their cooperative activities that result in some gain for all of them.

A group of k cooperating members is called coalition, s. All participating members form the grand coalition S that consists of all n participants, $k \leq n$. Each nonempty coalition has a value $V(s)$, which represents the cost (or the value of this coalition). The goal is to allocate the cost (or the total savings) of the grand coalition to each member in a "fair" way. The Shapley value provides a "fair" allocation in the sense that each member k is compensated proportionally to their marginal contribution $V(s) - V(s-k)$, which is then averaged over all possible different combinations in which the coalition can be formed.

The Shapley value, Sh_k, for each participant k is calculated as

$$Sh_k = \sum_{S \subset n} \frac{(s-1)!(n-s)!}{n!}[V(s) - V(s-k)],$$

where s is the number of participants in coalition S; summation is performed over all possible coalitions, which participant k joins; $(s-1)!$ is the number of arrangements for participants before joining s; $(n-s)!$ is the number of arrangements for participants after joining s; and $n!$ is the total number of all possible coalitions.

Table 6.2 Participants, costs, and coalitions for four-members saving allocation game

Participant (member)	Hospital {1}	Physician group {2}	SNF {3}	HHA {4}		
Median cost for stand-alone participants per episode-of-care	$15,000	$8,000	$6,000	$3,000		
Two-members coalitions	{12}	{13}	{14}	{23}	{24}	{34}
Cost reduction % for two-members coalition vs. stand-alone member cost	7%	4%	6%	3%	3%	2%
Cost for two-member coalitions	$21,390	$20,160	$16,920	$13,580	$10,670	$8,820
Three-member coalitions	{123}	{124}	{134}	{234}		
Cost reduction % for three-member coalition vs. stand-alone member cost	11%	8%	10%	7%		
Cost for three-member coalitions	$25,810	$23,920	$21,600	$15,810		
Grand coalition				{1234}		
Cost for grand coalition (13% saving over noncooperating stand-alone providers)				$27,840		

Thus, the Shapley value is computed by calculating the average marginal contribution that participant k brings to a coalition (group) s if this participant joins any coalition, and all coalitions for this participant k are formed in random order (there is no way of taking advantage of a more beneficial coalition order).

Participants have incentives to stay voluntarily in the coalition if (1) the participant's coalition cost is lower than its stand-alone cost (or the saving is higher)-individual rationality; (2) any subgroups' costs are lower than that of combined stand-alone costs of the subgroups' participants (or the saving is higher)-subgroup rationality; and (3) the total costs (savings) must be completely distributed among all cooperating participants, i.e., no participant's share of the costs can be reduced without increasing another participant's share. If these conditions are met, then the so-called *core* of cost sharing is not empty. Sometimes, though, the *core* can be empty. This means that there is no unique cost allocation that satisfies all participants. In this case, unsatisfied participants can leave the cost sharing coalition. The Shapley value can always be calculated even if the *core* is empty. However, if the core is not empty, then the Shapley value satisfies all participants; they all have incentives to stay in the coalition and get their "fair" cost (savings) allocation.

In this section example there are $n = 4$ participants that can form six paired (two-member) coalitions, four triple (three-member) coalitions and one grand coalition of all four participants, as indicated in Table 6.2.

Table 6.3 The Shapley values and saving allocations for four-members game

Cost due to cooperation: Shapley value	Cost due to cooperation: Shapley value	Cost due to cooperation: Shapley value	Cost due to cooperation: Shapley value
{1}	{2}	{3}	{4}
$13,402	$7,000	$5,047	$2,392
Saving {1}	Saving {2}	Saving {3}	Saving {4}
$1,598	$1,000	$953	$608
Saving allocation % {1}	Saving allocation % {2}	Saving allocation % {3}	Saving allocation % {4}
10.7%	12.5%	15.9%	20.3%

The Shapley value formula for participant 1 contains eight marginal values terms for this participant: one single, three double, three triple, and the grand coalition that include participant 1, as indicated in Table 6.2 in figured brackets{}. These terms are summed up with coefficients $(1-1)!(4-1)!=6$, $(2-1)!(4-2)!=2$, $(3-1)!(4-3)!=2$, and $(4-1)!(4-4)!=6$, respectively. The sum is then divided by $n!=4!=24$. Similarly, the Shapley value formula for participant 2 contains eight terms with the same numerical coefficients for marginal values of single, double, triple, and grand coalitions that include participant 2. The Shapley values for participants 3 and 4 are calculated in the same way.

The Shapley values for all four participants, their savings, and percent of saving allocations are indicated in Table 6.3. It follows from this table that the *core* is not empty. All four participants have economic incentive to stay in the coalition because their stand-alone costs would be higher. The Shapley values allocate higher savings to the lower cost participants (SNF and HHA) than simple cost-proportional allocations. This will likely encourage their cooperation with the hospital and physicians in providing coordinated patient care rather than position themselves as stand-alone providers. At the same time, the Shapley values allocate somewhat lower savings to hospital and physicians, so that the total saving for all participants remains the same.

In summary, cost allocation problem often arises in many business situations that benefit from nonlinear effect of economy of scale (volume discounts) or cooperating partners. Examples are bundled payment to providers; pooling resources where savings from cooperation (aggregation) must be distributed "fairly" between the beneficiaries of pooling, such as distribution of surgical costs in the presence of patient queues (Gonzalez and Herrero 2004); joint hospitals' diagnostic imaging or surgical centers, etc. One of the best approaches to "fair" costs (gains) allocation in this type of problem is the Shapley value or its modifications developed in the framework of mathematical game theory. In fact, any system in which control decision should be made based on the sum of its parameters can benefit from this technique (Reinhardt and Dada 2005).

Chapter 7
Summary of Some Fundamental Management Engineering Principles

Abstract Some fundamental management engineering principles in healthcare settings are summarized. These principles have been illustrated both by examples presented in this book and in examples published elsewhere.

Keywords Management engineering principles • Decision-making

Some fundamental management engineering principles in healthcare settings are summarized below. These principles have been illustrated both by examples presented in this book and in examples published elsewhere (Kolker 2010, 2011). Knowledge and understanding of these fundamental principles alone would help decision-makers to steer in the right direction even without building detailed complex simulation operational models, or conducting a sophisticated multivariate data analysis. These general management principles play in healthcare settings a role of the laws of physics in natural sciences.

- For systems with a similar type of service, mutually interchangeable (pooled) resources are more efficient in terms of patient wait time and throughput than specialized (dedicated) resources with the same total capacity/workload (illustrated in Sects. 2.3.1.3 and 2.3.9.3).
- If specialized (dedicated) resources are needed due to patient privacy, infection control, nonmovable equipment, or other special factors, then some additional capacity should be planned and budgeted to cover the loss of resources' efficiency.
- Specialized resources (staff, operating or procedure rooms, beds, etc.) typically cost more than mutually interchangeable (pooled) resources (illustrated in Sects. 2.3.1.3, 2.3.9.2, and 6.1).
- Because of variability of patient arrivals and service time, a reserved capacity (sometimes up to 40%) is usually needed to avoid regular operational problems due to excessive wait time and long lines (illustrated in Sect. 2.3.5.3).

A. Kolker, *Healthcare Management Engineering: What Does This Fancy Term Really Mean?*, SpringerBriefs in Health Care Management and Economics, DOI 10.1007/978-1-4614-2068-2_7, © Alexander Kolker 2012

- The higher the degree of randomness in arrival rate and service time, the lower the unit operational performance in terms of patient queue size, patient wait time, and the unit utilization.
- Reduction of process variability is the key to patient flow improvement, increasing throughput and reducing delays (illustrated in Sect. 2.3.6.3).
- Size matters. Large hospitals (units) always have better operational performance characteristics (lower wait time and the number of patients in the queue, higher utilization) than small hospitals (units) with the same patient volume to size ratio (illustrated in Sect 2.3.7.2).
- Process improvement efforts based on simple linear proportional adjustments of input values and direct benchmarking could be misguided and short-lived if the scale effect (organization size) is not taken into account (illustrated in Sect. 2.3.7.2).
- Generally, the higher utilization level of the resource (good for the organization), the longer the wait times to get this resource (bad for patients). For random patient arrival and random service time, a utilization level higher than 80–85% typically results in significant increase in patient wait time (illustrated in Sects. 2.3.6.2 and 2.3.7.2).
- Workload leveling (smoothing) of elective scheduled procedures is an effective strategy for reducing wait time and improving patient flow (illustrated in Sect. 2.3.8.3).
- Improvement of separate subsystems or units/departments (local optimization or local improvement) does not necessarily result in the improvement of the entire hospital system; a system of local improvements could be a very inefficient system (illustrated in Sect. 2.3.11.2).
- Analysis of an entire complex system is usually incomplete and can be misleading if it does not take into account the *subsystems' interdependency* (illustrated in Sect. 2.3.11.2).
- Scheduling appointments (jobs) in the order of their increased duration variability (from lower to higher variability) results in a lower overall cycle time and patient wait time (illustrated in Sect. 2.4.1.2).
- In a series of dependent activities, only a bottleneck defines the throughput of the entire system. A bottleneck is a resource (or activity) whose capacity is less than or equal to demand placed on it (illustrated in Sect. 2.4.2.2).
- An unfulfilled service request backlog (appointments, discharges, document processing, etc.) can exist and remain stable even if the average high variability demand is less than service capacity (illustrated in Sect. 2.4.2.2).
- Capacity, staffing, and financial estimations based on average input values without taking into account the variability around the averages result in significant underestimation or, sometimes, overestimation of required resources (except for strictly linear relationship between the input and output). This is called the flaw (deception) of averages (illustrated in Sects. 2.3.1–2.3.10 and 2.4.1–2.4.5).

- For low disease prevalence (less than 30%), pooled screening specimen testing is more efficient than individual specimen testing (illustrated in Sect. 3.4.2).
- In order to generate forecast of a time series, only the most recent, strongly correlated past data-points should be used. Data-points that are weakly correlated to the newer data-points (or not correlated at all) should not be used for making the forecast (illustrated in Sects. 4.1.2 and 4.2.2).
- Identifying separate significant independent contributing factors (variables) from a large observational dataset that contains mutually correlated data is not possible. In order to determine the significant contributing factors, the original dataset should be decomposed into mutually uncorrelated components using principal components methodology or another appropriate advanced multivariate data analysis technique (illustrated in Sect. 5.1.2).
- Principal component decomposition and cluster analysis can be viewed as techniques for efficient data volume reduction in large observational datasets. Principal component decomposition reduces the number of variables (columns); cluster analysis reduces the number of observations (rows) of the variables (illustrated in Sect. 5.2.2).

Chapter 8
Concluding Remarks

Let us summarize the main points of this Brief Series book. The main goal of the book is helping to bridge the gap in mutual understanding and communication between management engineering professionals and hospital and clinic administrators empowered to make managerial decisions.

This book is intended primarily for hospital/clinic leadership to raise their understanding of the powerful management engineering methodology as an indispensible aid for quantitative decision-making in the current era of great demand for much more efficient use of available healthcare resources. At the same time, this book can serve as a concise text and compendium of introductory problems/projects (with some extensions) for graduate students in Healthcare Management and Administration, as well as for MBA programs with an emphasis in Healthcare.

Toward this goal, an overview of the domain of healthcare management engineering is provided. The focus of this book is the application of management engineering principles and methodologies on the scale of a separate stand-alone hospital or a large clinic.

It is emphasized in this book that traditional managerial decision-making is based on past experience, feelings, intuition, simple linear projections, or calculations based on the average input values. Thus, the traditional approach does not have a proper means to take into account the inevitable process variability, uncertainty, scale, and interconnections that are critical for making sustainable and justified managerial decisions for real hospital operations.

On top of that, there is a general human tendency to avoid the complications of incorporating uncertainty in decision-making by ignoring it or turning it into artificial certainty. It is illustrated in this book why such a practice often results in highly inaccurate and short-lived outcomes made by traditional management decision-making.

In contrast, healthcare management engineering can broadly be defined as a systematic way of developing managerial decisions for efficient allocating of material, human, and financial resources needed for delivery of high quality care using analytic mathematical and computer simulation methods. Management decisions for

A. Kolker, *Healthcare Management Engineering: What Does This Fancy Term Really Mean?*, SpringerBriefs in Health Care Management and Economics, DOI 10.1007/978-1-4614-2068-2_8, © Alexander Kolker 2012

leveraging resources that best meet system performance objectives are based on comparative analysis of validated mathematical and computer simulation models.

Management engineering methodology is capable of revealing the deep interconnections between the elements of such a complex system as a hospital or a large clinic; it naturally takes into account the economy of scale. Management engineering is indispensable in understanding responses of the processes and systems to different inputs with random and nonrandom variability. This understanding makes it possible, in turn, to predict performance and/or real resource requirements, allowing decision-makers to be proactive rather than reactive.

No concept or methodology can truly be convincing without multiple concrete and practically relevant examples of its application. Therefore, in this book traditional managerial decision-making and management engineering methodology are applied side by side to analyze 26 concrete operational management problems adapted from hospital and clinic practice. The problem types include: clinic, bed, and operating rooms capacity; patient flow; staffing and scheduling; resource allocation and optimization; forecasting of patient volumes and seasonal variability; business intelligence and data mining; and game theory application for allocating cost savings between cooperating participants.

Detailed examples of applications are provided for such powerful techniques as discrete event simulation, queuing analytic theory, linear and probabilistic optimization, forecasting of a time series, principal component decomposition of a large observational dataset and cluster analysis, and the Shapley values for "fair" cost (savings) allocation between cooperating providers of care working toward a common goal.

A summary of some fundamental management engineering principles is provided. Knowledge and understanding of these fundamental principles alone would help the decision-makers to steer in the right direction even without building detailed operational models. These principles, which play a role of laws of physics in natural sciences, will also greatly help in developing sorely needed mutual understanding between management engineering professionals and hospital administrators.

The author sincerely hopes that the more healthcare management administrators become familiar and comfortable with quantitative decision-making methodology offered by management engineering, the more they will become passionate and supportive in its widespread practical application.

Appendix

Summary of some quantitative methods used for the various management engineering applications:

- Data envelopment analysis—to compare relative efficiency of units/ organizations
- Decision trees—to perform probabilistic analysis of different options
- Discrete event simulation modeling—widely used for numerous applications. Illustrated in multiple examples in Sects. 2.3 and 2.4
- Forecasting a time series (linear and/or nonlinear), as well as recursive forecasting technique—illustrated in Sects. 4.1 and 4.2
- Graph theory—for facilities layout development, facilities location, and ambulance routings
- Integer programming—some or all decision variables are only integer numbers
- Just-in time (dynamic pull system—based on actual usage)
- Linear programming-optimization of linear systems subject to a linear set of constraints—illustrated in Sects. 3.1 and 3.2
- Nonlinear optimization problem (optimization function and/or constraints are nonlinear)
- Stochastic programming (linear, integer, mixed-integer, nonlinear with random parameters, and variables)
- Markov processes (chains of probabilistic events)
- Materials requirement planning and inventory control–optimal order quantity
- Monte-Carlo simulation (static probabilistic analysis)
- Network analysis (critical path—for total duration of interrelated activities)
- Queuing analytic theory—illustrated in Sects. 2.3.1–2.3.3
- Regression analysis with principal components for multiple mutually correlated factors—illustrated in Sect. 5.1
- Sampling quality control
- Theory of constraints: the concept of a bottleneck in a series of interdependent events (activities)—illustrated in Sects. 2.3.11 and 2.4

A. Kolker, *Healthcare Management Engineering: What Does This Fancy Term Really Mean?*, SpringerBriefs in Health Care Management and Economics, DOI 10.1007/978-1-4614-2068-2, © Alexander Kolker 2012

- System dynamics: methodology for macrolevel simulation of very large distributed systems such as, for example, a nationwide system of hospitals or care delivery organizations
- Cooperative game theory and its application for costs allocation—illustrated in Sect. 6.1

Note: material and views presented in this book are developed solely by the author. They do not represent in any way the views of the current author's affiliation, Children's Hospital and Health System, Wisconsin, USA, nor management engineering work performed in this organization.

References

Abraham, G., Byrnes, G.,Bain, C. 2007. *Short-Term Forecasting of Emergency Inpatient Flow.* IEEE Transactions on Information Technology in Biomedicine, IEEE TITB-00211-2007.R1, pp. 1–9

Abu-Taieh, E., El Sheikh, A R., 2007. *Commercial Simulation Packages: A Comparative Study.* International Journal of Simulation, 8 (2), 66–76

Armstrong, W., Taege, A., 2007. *HIV screening for All: the New Standard of Care.* Cleveland Clinic Journal of Medicine. 74 (4), pp. 297–301

Ben-Arieh, D., Wu, Chih-Hang, 2011. *Reducing Patient Waiting Time at an Ambulatory Surgical Center.* Chapter 12. In: Kolker, A.,& Story, P. (Eds). Management Engineering for Effective Healthcare Delivery: Principles and Applications (pp. 1–512). IGI-Global, doi:10.4018/978-1-60960-872-9

Berwick, D., 2011. *Observations on Initiating Systems Change in Healthcare: Challenges to Overcome.* In: Engineering a Learning Healthcare System: A Look at the Future. Workshop Summary. Institute of Medicine (IOM), Washington, DC. The National Academies Press. p. 58 http://www.nap.edu/catalog.php?record_id=12213

Blum, A., Shea, S., Czeisler, C., Landrigan, C., Leape, L., 2011. *Implementing the 2009 IOM Recommendations on Resident Physician Work Hours, Supervision and Safety. White Paper. Dovepress.* Nature and Science of Sleep, (3): 47–85. http://dx.doi.org/10.2147/NSS.S19649

Bozzette, S., 2005. *Routine Screening for HIV infection-timely and cost effective.* New England Journal of Medicine, 352, pp. 620–621

Butler, T., 1995. *Management Science/ Operations Research Projects in Health Care: The Administrator's Perspective.* Health Care Management Review, 20 (1):19–25

Buttell Crane, A., 2007. *Management Engineers.* Hospitals & Health Networks (H&HN), Management/Governance, April, 50. http://www.hhnmag.com/hhnmag_app/jsp/articledisplay.jsp?dcrpath=HHNMAG/Article/data/04APR2007/0704HHN_FEA_Management&domain=HHNMAG

Carter, M., 2002. *Health Care Management. Diagnosis: Mismanagement of Resources.* Operation Research/Management Science (OR/MS) Today. April, 29: 2, pp. 26–32

Cayirli, T., Veral, E., Rosen, H., 2006. *Designing Appointment Scheduling Systems for Ambulatory Care Services.* Health Care Management Science, 9; 47–58

CDC report, 2008. *HIV Prevalence Estimates-USA, 2006.* MMWR weekly, 57(39), pp. 1073–1076

Compton, W. Dale, Reid, P., 2008 (Spring). *Engineering and Health Care Delivery System (Editorial).* The Bridge, 38 (1), 3–5 (National Academy of Engineering). http://www.nae.edu/File.aspx?id=7417

Costa, A., Ridley, S., Shahani, A., Harper, P, De Senna, V.,Nielsen, M., 2003. *Mathematical Modeling and Simulation for Planning Critical Care Capacity.* Anesthesia, 58, pp. 320–327

D'Alesandro, J., (2008). *Queuing Theory Misplaced in Hospitals*. Management News from the Front, Process Improvement. PHLO. Posted Feb 19, 2008 at http://phlo.typepad.com

De Bruin A., van Rossum, A., Visser, M., Koole, G., 2007. *Modeling the Emergency cardiac in-patient flow: an application of queuing theory*. Health Care Management Science, *10*, pp. *125–137*

Fabri, P., 2008. *Can Health Care Engineering Fix Health Care?* American Medical Association Journal of Ethics, Virtual Mentor, *10(5), pp. 317–319*

Fowler, J., Benneyan, J., Carayon, P., Denton, B., Keskinocak, P., & Runger, G., 2011. *An Introduction to a New Journal for Healthcare Systems Engineering*. IIE Transactions on Healthcare Systems Engineering, 1, pp. 1–5

Fuhs, P., Martin, J., Hancock, W., 1979. *The Use of Length of Stay Distributions to predict Hospital Discharges*. Medical Care, XV11, 4, p. 355–368

Gallivan, S., Utley, M., Treasure, T., Valencia, O., 2002. *Booked inpatient admissions and hospital capacity: mathematical modeling study*. British Medical Journal, 324, 280–282

Glantz, S., Slinker, B., 2001. Applied Regression & Analysis of Variance. 2-nd Ed., McGraw-Hill, Inc

Goldratt, E., Cox, J., 2004. The Goal. 3-rd Ed., North River Press, Great Barrington, MA, p. 384

Gonzalez, P., Herrero, C., 2004. *Optimal Sharing of Surgical Costs in the Presence of Queues*. Mathematical Methods of Operations Research. 59: 435–446

Green, L. 2006. *Queuing Analysis in Healthcare*. In Hall, R. (Ed.). Patient Flow: Reducing Delay in Healthcare Delivery (pp. 281–307). Springer, NY

Green, L., 2004. *Capacity Planning and Management in Hospitals*. In M. Brandeau, F. Sainfort, W. Pierskala, (Eds.), *Operations Research and Health Care. A Handbook of Methods and Applications*, (pp. 15–41). Boston: Kluwer Academic Publisher

Hall, R. 1990. *Queuing methods for Service and Manufacturing*. Upper Saddle River, NJ: Prentice Hall.

Hand, D., Mannila, H., Smyth, P., 2001. *Principles of Data Mining*. MIT Press, pp. 546.

Haraden, C., Nolan, T., Litvak, E., 2003. *Optimizing Patient Flow: Moving Patients Smoothly Through Acute Care Setting*. Institute for Healthcare Improvement Innovation Series 2003. White papers 2, Cambridge, MA

Harrison, G., Shafer, A., Mackay, M., 2005. *Modeling Variability in Hospital Bed Occupancy*. Health Care Management Science, 8, 325–334

Hlupic, V., 2000. Simulation Software: A Survey of Academic and Industrial Users. *International Journal of Simulation, 1*(1), 1–11

Ingolfsson, A., & Gallop, F., 2003. *Queuing ToolPak 4.0*. Retrieved from http://apps.business.ualberta.ca/aingolfsson/qtp/

IHI, 2011. Institute for Healthcare Improvement (IHI). http://www.ihi.org/knowledge/Pages/Tools/AppointmentSequenceSimulation.aspx

IOM (Institute of Medicine), 2011. *Engineering a Learning Healthcare System: A Look at the Future*. Workshop Summary. Washington, DC. The National Academics Press. http://www.nap.edu/catalog.php?record_id=12213

Jobson, J.D., 1992. Applied Multivariate Data Analysis. V. 2. Categorical and Multivariate Methods. Springer-Verlag New-York, LLC

Joustra, P., van der Sluis, E., van Dijk, N., 2010. *To pool or not to pool in hospitals: a theoretical and practical comparison for a radiotherapy outpatient department*. Annals of Operations Research, 178: pp. 77–89

Joustra, P., de Witt, J., Van Dijk, N., Bakker, P., 2011. *How to juggle priorities? An interactive tool to provide quantitative support for strategic patient-mix decisions: an ophthalmology case*. Health Care Management Science, doi 10.1007/s10729-011-9168-5

Kall, P., Mayer, J. 2005. Stochastic Linear Programming. Models, Theory and Computation. Springer Science & Business Media, New York, NY, pp. 397

Klassen, K. J., Rohleder, T. R. 1996. *Scheduling Outpatient Appointment in a Dynamic Environment*. Journal of Operations Management, *14, pp. 83–101*

Kolker, A., 2008. *Process Modeling of Emergency Department Patient Flow: Effect of Patient Length of Stay on ED Diversion*. Journal of Medical Systems, *32(5), pp. 389 – 401*. http://dx.doi.org/10.1007/s10916-008-9144-x

Kolker, A., 2009. *Process Modeling of ICU Patient Flow: Effect of Daily Load Leveling of Elective Surgeries on ICU Diversion.* Journal of Medical Systems, *33(1), pp. 27–40* http://dx.doi.org/10.1007/s10916-008-9161-9

Kolker, A., 2010. *Queuing Theory and Discrete Events Simulation for Health Care.* In: Health Information Systems: Concepts, Methodologies, Tools, and Applications. vol. 4, Chapter 7.12. Joel Rodrigues (Ed.). IGI-press Global, 2010. pp. 1874–1915. http://www.irma-international.org/chapter/queuing-theory-discrete-events-simulation/49971/

Kolker, A., 2011. *Efficient Managerial Decision-Making in Healthcare Settings: Examples and Fundamental Principles. Chapter 1. In: Management Engineering for Effective Healthcare Delivery: Principles and Applications. A. Kolker, P. Story (Eds). IGI-press Global, pp. 1–45.*

Kopach-Konrad, R., Lawley, M., Criswell, M., Hasan, I., Chakraborty, S., Pekny, J., Doebbeling, B., 2007. Applying Systems Engineering Principles in Improving Health Care Delivery. Journal of General Internal Medicine, 22 (3), 431–437

Langabeer, J R., 2007. *Health Care Operations Management.* Jones and Bartlett Publishers, Sudbury, MA, pp. 438

Lawrence, D., 2010. *Healthcare: How Did We Get Here and Where Are We Going?* In: Handbook of Healthcare Delivery Systems. Chapter 1. Yih Yuehwern (Ed.), Boca Raton, FL, CRC Press, pp. 1.1–1.15

Lawrence, J., Pasternak, B., 2002. *Applied Management Science: Modeling, Spreadsheet Analysis, and Communication for Decision Making.* 2-nd Ed., Hoboken, NJ: John Wiley & Sons.

Leapfrog Survey Group, 2011. Proposed Changes to the 2011 Leapfrog Hospital Survey. http://www.leapfroggroup.org/media/file/SmPtSkedQuestions.pdf

Littig, S., Isken, M., (2007). *Short Term Hospital Occupancy Prediction.* Health Care Management Science, 10, p. 47–66

Litvak, E., 2007. A new Rx for crowded hospitals: Math. Operation management expert brings queuing theory to health care. *American College of Physicians-Internal Medicine-Doctors for Adults,* December 2007, ACP Hospitalist.

Litvak, E., Long, M., 2000. *Cost and Quality Under Managed Care: Irreconcilable Difference ?* The American Journal of Managed Care, *6(3), pp. 305–312*

Mayhew, L., Smith, D. 2008. *Using queuing theory to analyze the Government's 4-h completion time target in Accident and Emergency departments.* Health Care Management Science, 11, 11–21

Mayo Clinic, 2011. *Mayo Clinic Launches New Center to Focus on How Healthcare is Delivered.* Mayo Clinic News, January, 25. www.mayoclinic.org/news2011-rst/6151.html

Marshall, A., Vasilakis, C., El-Darzi, E., 2005. *Length of stay-based Patient Flow Models: Recent Developments and Future Directions.* Health Care Management Science, *8, p. 213–220*

McLaughlin, D., Hays, J., 2008. Healthcare Operations Management. Health Administration Press, Chicago, AUPHA Press, Washington, DC, pp. 466

McManus, M., Long, M., Cooper, A., Litvak, E., 2004. *Queuing theory accurately models the need for critical care resources.* Anesthesiology. *100(5), pp. 1271–1276*

McManus, M., Long, M., Cooper, A., Mandell, J., Berwick, D., Pagano, M., Litvak, E., 2003. *Variability in surgical caseload and access to intensive care services.* Anesthesiology. *98(6), pp. 1491–1496*

Moeller, D., Evans, J., 2010. *Episode-of-care payment creates clinical advantages.* Managed Care, 19 (1): 42–5

Motwani, J., Klein, D., Harowitz, R., 1996. *The Theory of Constraints in Services: part 2-examples from health care.* Managing Service Quality, *6(2),* pp. 30–34

Nikoukaran J., (1999). Software selection for simulation in manufacturing: A review. *Simulation Practice and Theory,* 7(1), 1–14

Ozcan, Y., 2009. *Quantitative Methods in Health Care Management.* Second Edition. Jossey-Bass. A Wiley Imprint, San Francisco, CA, pp. 438

Peek, N., 2010. *Data Mining.* Chapter 24. In: Yih, Yuehwern (Editor). Handbook of Healthcare Delivery Systems. CRC Press, Boca Raton, FL

PHLO, 2008. Creating and Running Proactive Hospital Operations website, http://phlo.typepad.com/phlo/2008/01/the-worst-thing.html

Prekopa, A., 1995. Stochastic Programming. Kluwer Academic Publisher. The Netherlands.

Press, W., Flannery, B., Teukolsky, S., Vetterling, W., 1988. *Numerical Recipes in C. The Art of Scientific Computing.* Cambridge University Press. New York, NY, pp. 735

Reid, P., Compton, W, Grossman, J., Fanjiang, G., Eds., 2005. *Building A Better Delivery System: A new Engineering / Healthcare Partnership.* National Academy of Engineering and Institute of Medicine. Washington, DC. The National Academy Press

Reinhardt, G., Dada, M., 2005. *Allocating the Gains from Resource Pooling with the Shapley Value.* Journal of the Operational Research Society, 56, pp. 997–1000

Ryckman, F., Yelton, P., Anneken, A., Kissling, P., Schoettker,P., Kotagal, U., 2009. *Redesigning Intensive Care Unit Flow Using Variability Management to Improve Access and Safety.* The Joint Commission Journal on Quality and Patient Safety, 35 (11), pp. 535–543

Saraniti, B., 2006. *Optimal Pooled Testing.* Health Care Management Science, 9, pp. 143–149

Savage, S, 2009. *The Flaw of Averages.* John Wiley & Sons, Inc, Hoboken, New Jersey, pp. 392

Savin, S., 2006. *Managing Patient Appointments in Primary Care.* In: Patient Flow: Reducing Delay in Healthcare Delivery. Chapter 5. Hall, R.W. (Ed). Springer, New York, NY. pp. 123–150

Schuster, HG. 1998. *Deterministic Chaos: Introduction.* Weinheim; Physik Verlag

Seelen, L., Tijms, H., Van Hoorn, M., 1985. *Tables for multi-server queues* New-York, Elsevier, pp. 1–449.

Sen, S., Higle, J., 1999. *An Introductory Tutorial on Stochastic Linear Programming Models.* Interfaces, 29, 2, pp. 33–61

Story, P., 2009. *Are We Thinking Systems Yet ?* American Society for Quality (ASQ), January, 2009. http://www.asq.org/pdf/healthcare/are-we-thinking-systems-yet.pdf

Swain, J., 2007. *Biennial Survey of discreet-event simulation software tools.* OR/MS Today, 34(5), October

Teow, K L., 2009. *Practical Operations Research Applications for Healthcare Managers.* Annals Academy of Medicine Singapore, *38 (6), pp. 564–566*

Tibshirani, R., Walther, G., Hastie, T., 2001. *Estimating the Number of Clusters in a Dataset via the Gap Statistic.* Journal of Royal Statistical Society, B. 63: 411–423

Tseng, C-L., Brimacombe, M., Xie, M., Rajan, M., Wang, H., Kolassa, J., Crystal, S., Chen, T., Pogach, L., Safford, M., 2005. *Seasonal Patterns in Monthly Hemoglobin A1C Values.* American Journal of Epidemiology, 161 (6): 565–574

Ulmer, C., Wolman, D., Johns,M., Eds. 2009. *Resident Duty Hours: Enhancing Sleep, Supervision, and Safety.* Institute of Medicine. Washington, DC: National Academic Press. http://www.iom.edu/Reports/2008/Resident-Duty-Hours-Enhancing-Sleep-Supervision-and-Safety.aspx

Valdez, R.S., Brennan, P.F. 2009. *Industrial and System Engineering and Health Care: Critical Areas of Research.* Background Report. University of Wisconsin-Madison, Madison, WI. Agency for Healthcare Research and Quality (AHRQ) publication, #09-0094-EF, Rockville, MD. http://healthhit.ahrq.gov/engineeringhealthfinalreport (#10-0079-EF, May 2010)

Vissers, J M.H. 1998. *Health Care Management Modeling: A Process Perspective.* Health Care Management Science, 1: 77–85

Winston, W. 2004. *Microsoft Excel Data and Business Modeling.* Microsoft Learning, pp. 624

Weber, D. O., 2006. Queue Fever: Part 1 and Part 2. May 10. *Hospitals & Health Networks, Health Forum.* http://www.hhnmag.com/hhnmag_app/jsp/articledisplay.jsp?dcrpath=HHNMAG/PubsNewsArticle/data/2006May/060509HHN_Online_Weber&domain=HHNMAG

Wullink, G., Van Houdenhoven, M., Hans, E., van Oostrum, J., van der Lans, M., Kazemier, G., 2007. *Closing Emergency Operating Rooms Improves Efficiency.* Journal of Medical Systems, 31: 543–546

Young, H.P. 1994. *Cost Allocations.* In: Handbook of Game Theory with Economic Application. V. 2, Aumann, R., Hart, S., (Ed), 1193–1230. Elsevier Science B.V.

Yoo, I., Alafaireet, P., Marinov, M., Pena-Hernandez, K., Gopidi, R, Chang, J-F., Hua, L., 2011. *Data Mining in Healthcare and Biomedicine: A Survey of the Literature.* J. Med. Systems, Published Online, May 3, 2011. DOI 10.1007/s10916-011-9710-5

Definitions of Key Terms

Operations research The discipline of applying mathematical models of complex systems with random variability aimed at developing justified operational business decisions.

Management science A quantitative methodology for assigning (managing) available material assets and human resources to achieve the operational goals of the system. Based on operations research.

Complex system A system that exhibits a mutual interdependency of components and for which a change in the input parameter(s) can result in a nonproportional large or small change of the system output.

Discrete event simulation One of the most powerful methodologies of using computer models of the real systems to analyze their performance by tracking system changes (events) at discrete moments of time.

Queuing theory Mathematical methods for analyzing the properties of waiting lines (queues) in simple systems without interdependency. Typically uses analytic formulas that must meet some rather stringent assumptions to be valid.

Simulation package Also known as a simulation environment. A software with user interface used for building and processing discrete event simulation models.

Flow bottleneck/constraint A resource (material or human) whose capacity is less than or equal to demand for its use.

Principal component methodology A methodology that allows one to perform a multivariate correlation analysis and identifies redundant original variables that carry little or no information while retaining only a few mutually uncorrelated new variables. These new variables called principal components; they are linear combinations of the original variables.

Printed by Publishers' Graphics LLC